UNIVERSITY OF NOTTINGHAM
10 0711435 9
WITHDRAWN
FROM THE LIBRARY

D1766789

# Towards an Ecology of Tectonics

# Towards an Ecology of Tectonics

## *The Need for Rethinking Construction in Architecture*



Editors:
Anne Beim
Ulrik Stylsvig Madsen





Authors:
Claus Bech-Danielsen
Charlotte Bundgaard
Johan Celsing
Karl Christiansen
Jonathan Hale
Bijoy Jain
Thomas Bo Jensen
David Leatherbarrow
Ulrik Stylsvig Madsen
Fredrik Nilsson
Ole Egholm Pedersen
Børre Skodvin
Peter Thorsen

**Edition Axel Menges**

# Content



2

1

## Cultural Ecology

## Industrialised Craft



# 3

## Construction Materialised

# Introduction
# An Ecology of Tectonics

**Anne Beim**
is professor of architecture and head of CINARK – Centre for Industrialised Architecture at The Royal Danish Academy of Fine Arts, School of Architecture (RDAFA) in Copenhagen. She has been chairing the cross-institutional research project "Towards a Tectonic Sustainable Building Practice" (2010-2014). Anne Beim holds a PhD in architecture from RDAFA and has conducted part of her PhD studies at the University of Pennsylvania School of Design. She is author/co-author of several books including: *Building the Future – Visions in Industrialised Housing 1970-2011* (2012), *Three Ways of Assembling a House* (2009), *Tectonic Visions in Architecture* (2004) and *Ecology and Architectural Quality* (2002).

An ecology of tectonics embeds the concept of buildings as parts tied together as a whole in a broader context of natural and cultural systems. This understanding feeds a new ethical dimension into tectonic practice that recognises the correlation between the materials used, the ecosystems they form a part of and the resources we share as common members of the global community. Using this sort of knowledge as a guiding principle in the design and construction of architecture seems crucial in view of the environmental crisis we are facing today.[1]

In an overall perspective, the aim of this book is to set the grounds for discussions about tectonic strategies as ways to approach the existing challenges in contemporary building practice, the point of departure being an ecological approach to the creation of architecture. Ecology is here defined in its widest sense, which besides ecological systems found in nature includes environmental dimensions, the life cycle of resources, social organisation, and the longevity of the contextual qualities of design.

The topic is approached from different angles represented by distinctive viewpoints, such as philosophical, historical and empirical, which also include cultural criticism, aesthetic analysis and reflections on architectural practice. Per se, the presented themes are vast in their scope – ranging from ontological studies of building designs, over geological and material investigations, to the craft of construction and industrialisation in architecture. Yet the themes closely focus on questions related to how architecture is constituted by means of physical matter, its making, and the fundamentals of resource thinking from a holistic perspective. Parallel to this, further questions address how these aspects of construction are culturally rooted as well as how various architecture-related cultures are formed by them.

**The need for rethinking construction in architecture.**

Due to the growing impact of climate change and predictions of natural resource scarcity it seems necessary to reconsider the basis on which architecture is conceived and being built.[2] How do we ensure a proper framework for creating a holistic, sustainable architecture that supports societal prosperity without compromising the living systems of the Earth, and by what means do we achieve building cultures that are more robust in view of future resource perspectives – building cultures that neither give up on human dimensions nor let go of the core qualities of empirical craft traditions?

In the years to come these questions will undoubtedly be central to political decision making and in the world of architecture, yet some may find it hard to recognize the correlation between a sustainable change for the better and the tectonic dimensions in architecture.

With reference to the questions raised, it is important to note that most of the prior definitions of tectonics and the making of buildings have been responses of profoundly different and much less complex circumstances in architectural construction than we know of today, both in terms of materials used, craft traditions and the evolutionary stage of construction technologies.[3] The present-day construction industry involves a multitude of materials (natural, artificial and composite), highly complex construction solutions, and industrialised building supplies produced by computer-aided manufacturing (CAD/CAM).[4]

Due to the generally expanding use of computers, architectural design phases are now closely linked to industrialised production and the construction phases of building, but also new demands for information and control in view of the fact that factors such as local climate conditions, performance of building physics and management of energy consumption are equally influencing the evolutionary course of the construction industry. Whereas the physical dimensions of the wide-ranging procedures linked to architectural construction can be defined as "primary technologies", the digital design tools and information systems can be characterised as a field of "secondary technologies" that are now included in the conception and making of buildings. As "technologies" they are more volatile, but at the same time they are situated in the real context of the building industry, comprising regulating systems based on international standards and certifications.[5]

As a result of this evolutionary track work, the contents and procedures of architectural practice have shifted the weight from "design" to "management" and architects now have the role of "technocrats" who spend their time processing flows of information that guide the assembly of complex technical building constructions and the needs of the management systems.[6] One could even go so far as to describe architects as "captive supporters"[7] of the very technologies they are subject to – as these technologies may happen to obstruct visionary improvements of architectural construction.

In other words, today's construction industry has developed into a paradox of autonomous interdependent systems largely ruled by global trends and economies. Within this context regional cultural identities and holistic approaches to architectural construction are difficult to maintain. Efforts to arrive at environmentally friendly solutions that are more comprehensive have likewise met with new barriers. In worst-case scenarios the result is a perfunctory collage architecture that no longer depends on local materials or empirical craft traditions – instead, other sorts of technologies and larger "systems of control" overshadow both the complexity and the particularity of the problems they try to solve.[8]

These sorts of dilemmas can be illustrated by one of the more critical consequences created by international political agendas concerning the reduction of energy consumption and GHG emissions. In a European context, the construction and operation of buildings is responsible for 40-50 percent of the total energy consumption, and the political objective is to reduce the use of energy by 20 percent by the year 2020.[9] Reaching this goal places high demands on the construction industry and built environment, but so far primary actions have focused on improving insulation standards and airtight construction almost in line with rigorous passive house standards.[10] Presently, we see construction solutions developed where excessive amounts of insulation material are being used, abolishing proper empirical construction solutions and building technologies. However, retrofitting the existing building stock according to present energy standards will not only affect the actual structures that have been built according to former building standards – but unintentionally will also change the appearance of historical building culture significantly.

In view of the heterogeneous character of environmental challenges, the "energy discourse" appears to be critical, as it does not imply a comprehensive (ecological) perspective. Furthermore, it tends to lead to somewhat dogmatic interpretations of the problems in question, emphasizing a mechanical understanding of technology – circumscribed by cause and effect. Despite the many good intentions and measurable effects that are linked to reducing fossil energy consumption, this "environmentally friendly" action happens to have adverse implications for the architectural design, building culture, construction practices, for building physics and indoor comfort, and for the way buildings may be able to adapt to various uses in a long-term perspective.

Thus, rethinking architectural construction in the age of environmental crises means rethinking how the use of material resources and the making of building can respond to a larger degree of complexity that includes a cultural (ethical) dimension and consequently implies rethinking how contemporary building technologies can be reconciled with the natural systems in which not only architecture, but our very lives are embedded.

**The idea of tectonics – contemporary theories of architectural construction?**

Architecture based on tectonic principles tells us the story of its making – it refers to its overall contextual setting and the embedded meaning – from the ideas forming the program, over particular construction details, to the weathering of the buildings in the course of time. In that sense it concerns the essence of construction and construing in architecture.[11] Tectonic thinking is hesitant to random construction solutions; it thus forms a critical resistance to shifting trends determined either by the economic interests of the building industry, paradoxes in building regulations – or by spontaneous architectural trends.[12]

Ideas and theories about material nature, how they perform as structural systems, and how they transform into construction design and architecture can be traced back to the very beginning of human civilisation.[13] The texts included in this book add to an established discourse of theories related to the making of buildings and the concept of tectonics - a field of knowledge that has been subjected to growing attention over the past couple of decades by architects in general. In order to understand how the presented texts relate to a larger landscape of ideas that respond to the evolutionary track of architectural construction, it seems relevant to describe a few main positions that can be identified across a couple of generations.

### The meaning of construction

If we look into architectural history, classical treatises of architecture have naturally described this field of knowledge, though only as a secondary subject matter considered by means of more formal systems, as refined classifications of orders, rules of proportions or as the divine composition of architectural elements.[14] Specific theories of architectural construction that can be defined in terms of the tectonic understanding we know today do not really appear until the middle of the nineteenth century. Seminal ideas published in the books *Die Tektonik der Hellenen*[15], by Karl Bötticher (1806-1889), *Der Stil in den technischen und tektonischen Künsten*[16] and *Die vier Elemente der Baukunst*[17] by Gottfried Semper (1803-1879) offered superior analyses of Greek architecture from antiquity (Bötticher), and vernacular building cultures and craft traditions (Semper), cultivating a new understanding of the constituting elements of architecture. They described this field by the Greek term: τεκτονικός and named it "the tectonic". (For more detailed accounts of its etymological and cultural roots see the chapter "An Etymology of Tectonics".)

A shared interest in both the rational and poetic aspects of materials and building construction characterised this first generation of "tectonic theoreticians". Also, they studied how variously the scope of origins and creation could be defined as an equal part of these matters, interpreting the tectonic as signifying a complete system binding all the parts of the building into a single whole.[18]

Karl Bötticher introduced the dualistic idea of *Kernform* (the form of the structural core) and *Kunstform* (the form of the artistic representation) in order to depict the basic relationship between material origins and idealized re-presentations of material properties and structuring forces.[19] He saw the tectonic realm as a source out of which material and structural innovations could emerge - to play a leading role in the formation of new constructions and building systems.[20] Semper was inspired by similar ideas, emphasizing materials as the essentials of the "act of becoming", where the tools and procedures used were considered equally important.[21] According to Semper "the tectonic" can be defined as a result of conscious artistic work. Following this argument tectonic design requires an artistic idea, which acts as an organizing and structuring principle - an overall principle that forms materials and constructions into coherent structure.[22]

### Tectonic culture

Important scholars such as Kenneth Frampton, Marco Frascari, Gevork Hartoonian and Mari Hvattum have in recent times offered significant contributions to the historical and philosophical field of tectonics and can be characterised as the second generation of tectonic thinkers. Building upon the founding ideas of Bötticher and Semper, they have elaborated the notion of the tectonic by examining complementary philosophical ideas and by comparing historical and contemporary construction practices.

In *Studies of Tectonic Culture: The Poetics of the Nineteenth and Twentieth Century Architecture*, Frampton analysed and described tectonic concepts and practices through a series of detailed case studies of contemporary architecture and constructions. Frampton apprehends Semper's distinction between technical and symbolic aspects of construction, referring to it as a dichotomy of the ontological and representational nature of building elements that brings about different cultural conditions. The fine essay "Tell-the-Tale Detail" by Frascari offers a different approach to the understanding of architectural construction by introducing the binomials *construction* and *construing*, whereby he brings the two dimensions - the actual and the formal - closer together.[23] In addition he identifies the tectonic detail as "the site for innovation and invention" developed through the playful "narrative" process of the architectural drawing and visionary thinking.

In *The Ontology of Construction*, Hartoonian examines tectonic in relation to Martin Heidegger's philosophical theories of technology and the dualistic nature of the classical Greek concept of *techné*, which unifies the means and the ends - the work and the meaning.[24] Also he explains the hermeneutical dimension of Semper's comprehension of "the new", which separated him from the usual modernist thinking that rejects history; "as ... [Semper] witnessed the disappearance of traditional forms of art, this made him formulate a theory that integrated ur-forms with new techniques and materials."[25]

Finally, in the historical thesis *Gottfried Semper and the Problem of Historicism*, Mari Hvattum describes Semper's understanding of the poetic (*poesis*) in a classical sense close to the Aristotelian of *praxis* as all man's relations and connections to the world - rejecting it as formal composition, but instead defining it by its ethical significance and its power to embody human situations.[26] As part of this understanding she elaborates on Semper's notion of making as "a particular mode of making informed by a particular mode of knowledge".[27]

### Digital tectonics

The radical changes in contemporary construction and manufacturing – and in architectural design processes provided by the use of computers – have led to a new paradigm of thinking in architectural culture referred to as *digital tectonics*. Here the books *Digital Tectonics* by Leach, Turnbull & Williams, *Atlas of Novel Tectonics* by Reiser + Umemoto, and *Refabricating Architecture* by Kieran & Timberlake have contributed to this third generation of tectonic theory.[28]

Leach, Turnbull & Williams argue that the original dichotomy between the immaterial digital system level and the tectonic physical level is about to disappear due to the emergence of "digital technologies". They believe this new way of working represents a "structural turn" in architectural culture. By reference to a number of building projects and digital experiments they show how it has sparked a new spirit for collaboration between architecture and engineering that seems to influence the making of buildings in the future.[29] At a very abstract level Reiser + Umemoto demonstrate a rigorous academic approach to the field of digital tectonics, classifying different design problems into three main categories: geometry, matter and operating. Through analysis by means of digital test models they have developed a set of design strategies to be used primarily as guidelines in their own design practice. Based on these inquiries Reiser + Umemoto come up with a new definition of the cultural dimension of materials: "Instead of seeing regionalism emerge out of distinct cultures as was once the case, a universal regionalism assumes a global culture in which materials logics engender regions rather than the other way around. In fact, territories are no longer only defined by physical locale."[30]

Through observations and examination of the contemporary building industry, Kieran & Timberlake offer a different approach to the field of architectural construction and design processes fostered by computerized tools, and they stress their communicative aspects and how these can be used for information management.

In that sense, they bring the discussion of contemporary manufacturing in the building industry and the role of the architect further than initially proposed by the advocates of digital tectonics. Kieran & Timberlake define today's architect as similar to the master builder of the past: a material scientist, product engineer, process engineer, user and client who creates architecture informed by commodity and art. By recognizing commodity as equivalent to art, they believe that architecture is made accessible, affordable and sustainable, as are the most exclusive consumer products available today. They claim: "[...] Modern humanism is communication not geometry!"[31]

### An ethical dialogue

As one summarises this brief review of ideas and theoretical positions, it is clear that, by different means, they represent an ongoing interest in "the nature of things". Defined by different terms throughout history, fundamental elements of building constitute central parts of the tectonic field: the material properties, the interaction between the physical dimension of construction and the creative force of construing, and the very technologies involved – tools, methods and ways to communicate. Questions about "origins" and authenticity have equally been posed as part of tectonic inquiries. Yet one of the most significant aspects referred to in recent theories is its cultural dimension, given by its ethical importance and "its power to embody human situations", as voiced by Hvattum.

In conclusion, it is important to note that the texts contained in this book are intended to offer an open dialogue that can feed into the next generation of tectonic theories. Through significant contributions by an international network of architectural scholars and practicing architects, the book provides a comprehensive representation of contemporary tectonic thinking and practice in architecture. Various topics are discussed under three major themes; "cultural ecology", "industrialised craft" and "construction materialised". These are further introduced in the beginning of the three parts of the book.

As a backdrop for the ever-changing challenges that are facing building construction and architectural practice it is important to add that tectonic thinking must continuously be reviewed. Therefore the collected texts look beyond a typical instrumental understanding of architectural construction and instead find resonance in ideas and projects that celebrate the complexity of architecture.

Anne Beim

Endnotes p. 208

# An Etymology of Tectonics

In dictionaries the concept of tectonics generally has two definitions:

- The movements of the Earth's surface in relation to the so-called tectonic plates of the lithosphere
- The design and internal structuring of a work of art, particularly in an architectural context

In the case of the first definition the use of the concept of tectonics is quite recent in so far as the knowledge of the movements of tectonic plates was developed and established as late as the mid-1900s.

The second definition, on the other hand, was linguistically constituted and conceptualised in Ancient Greece, for which reason an etymological study of the concept of tectonics must take its starting point in Ancient Greek (i.e., from 800 BCE).

If you trace it back directly, you find, time after time, that tectonics is connected with wood as a material. Woodworkers, including carpenters, joiners and ship's carpenters, are indiscriminately referred to by the term τέκτων. It is, however, a mistake, quite a common one actually, to connect the term specifically with only one specific material. A craftsman working with horn, for instance, is called Κεραζόος τέκτων; but also a poet, even a schemer, falls – albeit peripherally – under the category of tectonics.

Regarding the connection between the concept of tectonics and wood it is worth mentioning that Aristotle studied, in what we today know as Aristotle's Metaphysics, among other things, the basic nature of things. The aim of the study was to explain the fact that the coming into existence of a thing presupposes something that exists before it on which the thing is based. Aristotle, however, simply lacked a word to describe this thing. He knew that it was something which is given a different shape, a shape which is the result of the processing of......well, the thing he did not have a word for. In the absence of something better, instead of "inventing" a new word, he used what was to him closest at hand, and therefore most adequate, namely ύλη (*hyle*).

Today, ύλη refers to *substance*, *matter* – to *material*, actually. This is Aristotle's work. Before him ύλη meant *wood*, *forest*, *building timber*, *firewood* or simply *timber*. It is without doubt this original meaning which has coloured, or tainted, our understanding of tectonics.

Τεκτονικός and Τέκτων are the closest we get to a direct translation of tectonics. In general, the concept to a much greater extent indicates a method rather than any material aspect, namely the process of producing, manufacturing, processing, bringing about, preparing and causing, even breeding. τίκτω thus balances between poetry and the breeding of living creatures. The latter definition, however, distances itself from tectonics, whereas the first definition, poetry, definitely does not. The concept of tectonics aims at bringing forth something – although not just any process of bringing forth. In the tectonic context the process of bringing forth is the result of a special knowledge, applied in a particularly clever, skilful, cunning and crafty way. A process is initiated in a way that goes beyond what is customary, applying an elevated awareness and understanding. It refers to a creative process of a particularly high status: the work of a skilled carpenter, a master, a master builder, an artist. That is, someone who constructs, assembles or processes his material in such a way that something which surpasses the specifically pragmatic is created. Τέχνη (*techne*) occurs again and again in the context of tectonics, indicating that it is not just a question of craftsmanship, but also of elevated insightful, learning and artistic abilities. Tectonics overlaps aspects of poetry, and is thus not restricted *exclusively* to the process of bringing forth, but also includes creation in absolute terms.

Furthermore, tectonics points towards building construction. Τεκτοσύνη means the art of building/architecture. Originally it was clear that tectonics was particularly concerned with architecture, moreover with architecture understood as art. The language of the Ancient Greeks continuously and naturally developed over time, transforming into Modern Greek in 1453 CE.

If you look up tectonics in a dictionary of Modern Greek you will find τεκτονικός, for which two different interpretations are provided:

geologic __ tectonic
(μασονία) __ *freemason*

I have already discussed the first instance. The latter, however, is immediately surprising, and then again, perhaps not!

Freemasonry is based on the process of creating something. Furthermore, it is not just about any kind of creation, but about a special creative process which is based on and continues almost all the terms we have just seen were defined in antiquity. The fact that freemasons created their works of architecture in stone simply supports the argument that tectonics is not restricted to a specific material. On the contrary – the focus is, once again, on the circumstances.

Today, freemasonry is by many perceived as esoteric, verging on the mystical. But freemasonry is actually a doctrine that has its origins in ancient times, as a doctrine that provided a philosophical and psychological basis for alchemy, which is, however, seen as even more mystical – as something related to sorcery. What both alchemy and freemasons have in common, however, is the aspect of refinement. In the case of alchemy, the lead which is refined into gold is, of course, metaphoric. In reality the process of refinement is about civilising primitive, unrefined man (In Danish *unrefined man* might translate as *plumb man,* derived from Latin *plumbum* (=*lead*)), by gathering, developing and accumulating knowledge. This was the basis of the work of the freemasons: to work with yourself and other like-minded people on the philosophical and psychological level, in order to achieve a deep understanding of what it means to be a human being.

In Greek, *freemason* translates as μασονία, which shares some similarities with the English term *mason*. The term *freemason* arose in the Middle Ages as a way of describing master builders, stonemasons and architects. The work of a stonemason does precisely consist in refining a raw block of natural rock, transforming it into a refined cultural building element. But thorough the same process, the stonemason is also refined, as he learns from his work, which is continuously elevated to a still higher level. In this way a solid connection is made to the understanding of τέχνη (the concept of *techne*) and thus to the arts in the very highest meaning of the word. The knowledge acquisition and continuous development which made possible the creation of some of the greatest works of art in architectural history, the Gothic cathedrals, was practiced by freemasons as a consequence of the fact that they, in contrast to other craftsmen who were geographically tied to guilds, had liberated themselves and were consequently able to travel and gather knowledge from near and far. They simply became more knowledgeable than their geographically restricted colleagues. This acquired knowledge the freemasons developed in closed meetings, through discussions which included not only, although they were based on, specific building-related problems, but also metaphysical epistemological human aspects. This systematic scientific work, which has developed construction and the art of building to an ever higher level, has also continued and innovated the very way this development takes place. The freemasons truly worked in a tectonic manner. In Modern Greek the order of the freemasons is called τεκτονισμός.

In our etymological study of the concept of tectonics, τέχνη naturally takes a central role.

Τέχνη is the origin of the technology of our time – but that is not all. The concept, which covers a far wider territory, is dualistic, and so, from an overall perspective, includes instrumental as well as poetic aspects. The Ancient Greeks could not have imagined that it would be divided into a technical aspect and an artistic aspect. The latter came along with the first – noncausally. However, in the meantime, the concept has developed and made thinking in terms of just such a separation possible. In Modern Greek the arts have patented τέχνη, whereas technique has been banished to the term τεχνική and is understood as a rather technocratic matter.

In an architectural context, however, the concept of tectonics picks up the pieces after this linguistic slip.

Tectonics is a distilled product, which cannot be further distilled. The distillate is a fusion of the specific form, the specific material and the technique, which, through the processing of the likewise specific material, has created the aforementioned specific form. If the distillate is also poetry – which it incidentally created – it may be architecture!

Karl Christiansen



Endnotes p. 209

## Cultural Ecology

*Can architecture evoke ecological awareness in the cultural and social context of our everyday lives?*

# Introduction

*An ecological understanding of the world links the well-being of the individual to the conditions of its surrounding environment. How can this understanding form the basis of an architectural practice that integrates ecological principles in order to construct a meaningful cultural framework of everyday life?*

*Could this way of thinking lead to a new kind of cultural understanding and code of conduct in the building industry based on ecological values?*

In light of the current ecological crisis these questions seem to be unavoidable. We have for the last couple of centuries been mismanaging the natural resources to a degree that now threatens the well-being of future generations. A rethinking of the ethical dimension in relation to our surrounding environment seems essential to restore the balance of the natural ecosystems.

Culture and nature are often described as antitheses. The evolution from being "animals" or prehistoric creatures to becoming cultivated human beings introduced a growing distinction between nature and the human sphere – between our biological origins and dependency and our construction of artifice – i.e., various technologies. As humans we are both an integrated part of nature and detached from it due to our ability to conceive, process and transform nature. Our approach to the natural environment reflects a cultural perception and interpretation of its potentials. In other words, we construct a conceptualisation of nature, blurring the fact that we are an integral part of it.

An ecological way of thinking bridges this gap between nature and culture by focusing on the interdependency between the individual (human or animal) and its surrounding environment. This introduces an ethical dimension in our relation to nature. When interacting with nature's ecosystems we do not only affect the organisation and balance of the systems, but our interference also has a direct effect on our own situations by determining our future access to the resources of the systems.

It is important to stress that the question is not how much we can exploit the natural resources without catastrophic implications for our own situation but rather how we can secure the natural balances of the ecosystems and thereby their capacity for regenerating. In other words, the question of resilience in regard to our ecological systems becomes the pivot. We need to concede that we are small parts of a larger whole, and thereby accept the fact that we are responsible for the well-being of our natural surroundings in general – including in a long-term perspective.

This way of thinking needs to form an integrated part of our cultural practices. Ecological considerations and values must be deeply embedded in the way we structure and manage our physical surroundings, and restructuring the way we think and practice architecture plays an important role in this process. In order to reach these goals we need to define new tectonic strategies that embed ecological awareness in the cultural and social context of our everyday lives.

**Positions and Perspectives**

This part of the anthology introduces four different perspectives on how to approach the concept of cultural ecology. The first two texts, by David Leatherbarrow and Karl Christiansen, offer a theoretical discussion presenting two different positions in the understanding of the relationship between nature and culture seen from a tectonic perspective. Sustainability seen in the light of the development of modernist architecture is the focus point of the third text, by Claus Bech-Danielsen, introducing tectonics as a way to bridge our conception of modernity with the ecological challenges we are facing today. The final contribution to the discussion of cultural ecology by Børre Skodvin is seen in light of the practicing architect offering a deep insight into how tectonic strategies can inform a project and embed the construction both in a cultural, topographical and ecological context.

The four positions in the field can be recapitulated in these short descriptions of the different perspectives of the contributions to the discussion:

In his text "A Study in Cultural Ecology" the American architectural theoretician David Leatherbarrow introduces the concept of "cultural ecology" as a way to approach the current ecological crisis by realising the importance of both the environmental and cultural aspects of the creation of architecture. The selection, treatment and assembly of materials, together with the configuration and qualification of settings can be evidences of ecological order if cultural norms guide the orientations and involvements that define the design - in this way ecology becomes a mode of being in the world - both a place, a pattern and a way of life. The discussion is based on an explanation of the etymological and scientific roots of ecology and a careful study of the project Eco House designed by the Norwegian architect Sverre Fehn.

The Danish architect and theoretician Karl Christiansen advocates the thesis that there is a close correlation between the movements of the Earth's crust and the creation of architecture - conceptually as well as at the concrete level - in his text "Nature, Culture, Tectonics of Architecture". It can be seen as a profound investigation of the link between tectonics as both a geological and an architectural phenomenon. In both fields the tectonic form comes from the "inside". In geology the form of the Earth's crust is a result of the movements and tensions within the earth itself. In the field of architecture the tectonic form can be seen as one with the parameters that have created it - as a fusion of material, technique and form. The text is based on etymological studies and introduces an insight into the significance of working with tectonic principles as the basis of architectural practice.

Claus Bech-Danielsen is a researcher in the field of architecture in Denmark and focuses on sustainability as an important objective for current architectural development in his contribution "Sustainable Architecture: An Abstract Culture's Search for Concrete Roots". He recalls that the solutions of environmental and social issues were already important driving forces behind the development of modernist architecture. However, environmental problems have changed: The environmental problems that modernists reacted against could be experienced by the senses in cities and dwellings. This made it possible to link a solution of the problems with the development of sensual architectural qualities. Today, environmental problems are increasingly abstract - they develop far from the building site, and can be registered only by means of advanced testing equipment. This makes it complicated to link the solution of environmental problems with the development of architectural quality. The article discusses tectonics as a way to overcome this problem.

The unpredictable limitations and opportunities occurring during the building process have an important impact on the architectural expression of the final building that architect Børre Skodvin from the Norwegian architectural office Skodvin & Jensen underlines in his text "The Complexity of Realness". Understanding and defining the limits of potential compromises in this process plays an important role in the work of the office. The text gives a detailed description of the development of the Mortensrud Church project, offering a deep insight into the methods used by the architects to develop solutions in close interaction with the topographical and ecological context. The result of this process is a building with a complex and fragmented character that through its connection with the surrounding environment and the exposed construction tells us the story of its own making.

Ulrik Stylsvig Madsen

David Leatherbarrow

# A Study in Cultural Ecology

*Above all, we must remember that nothing that exists or comes into being, lasts, or passes, can be thought of as entirely isolated, entirely unadulterated. One thing is always permeated, accompanied, covered or enveloped by another; it produces effects and endures them. And when so many things work through one another, where are we to find the insight and discover what governs and what serves, what leads the way and what follows?*
J.W. Goethe, *Toward a Theory of Weather*

In our time it is virtually impossible to respond to any architectural task without considering its ecological implications. This is because we have come to realize we are in the midst of a global ecological crisis. While professors and students of my generation once flattered themselves with the idea that attention to ecological concerns was optional, all of today's students and many of their teachers know they have no choice. Today's student does not decide whether or not to respond to the crisis, but when to do so and how to begin.

Why the crisis and what might be done about it? The answers are many. Certainly population growth has something to do with it; likewise, concomitant scarcity of resources. Scarcity might not be such an issue if there were not corresponding problems of greed and waste: socially, in our habits of consumption and insistent desire to obtain and display wealth, and spatially, in our expansive settlement patterns, redundant technologies, etc. There are both empirical and ethical dimensions to the current ecological crisis.

Design's response to this state of affairs is generally assumed to involve problem solving. The first premise of the typical approach is that the current status quo of so-called "advanced" civilisations should be maintained, as if our desires were also entitlements, and that neither should be externally constrained. Second, we tend to think that the solutions we devise should be widely applied. This assumes they are essentially technical in nature, for one of the basic characteristics of the technical procedure is its independence from territorial obligations. Third, the epistemological basis for the development of

Le Corbusier, Poem of the Right Angle, B3 Esprit, 1947-53. © Foundation Le Corbusier these types of solutions is rational thinking. Ecology, then, is a science with an empirical basis, developed through rational methods, resulting in design and construction techniques that can be widely applied to maintain (maybe even enhance) the status quo. There is nothing wrong about this style of thought other than its partiality: from the point of view of design it tells half the story. The term introduced in my title – "cultural ecology" – is meant to indicate that the problem is actually two-part.[1]

To explain this rather uncommon term I would like to cite an example from its early usage. In 1955 the anthropologist Julian Steward wrote: "Although initially employed with reference to biotic assemblages, the concept of ecology has naturally been extended to include human beings, since they are part of the web of life in most parts of the world. Man enters the ecological sphere, however, not merely as another organism related to other organisms in terms of physical characteristics. He introduces the super-organic factor of culture, which also affects and is affected by the total web of life... The problem of explaining man's cultural behaviour is of a different order than that of explaining his biological evolution... cultural patterns cannot be analyzed in the same way as organic features." Steward concludes his opening definition with two additional points: "Culture, rather than genetic potential for adaptation, accommodation, and survival, explains the nature of human societies... [and] cultural ecological adaptations constitute creative processes." Ecological science is not wrong for architecture, just inadequate because partial.

*"Culture, rather than genetic potential for adaptation, accommodation, and survival, explains the nature of human societies [and] cultural ecological adaptations constitute creative processes"*

Historically speaking, the techno-scientific sense of ecology is rather new. Ecological science emerged in the 19th century thanks to the pioneers of the new biology, Ernst Haeckel (who coined the term) and Charles Darwin. Earlier Carl Linnaeus had used the phrase "economy of nature" in the same sense. The "eco" of ecology is also the root of the word economy, which points to the tension between spending and saving – also the possibilities of sharing. Darwin's debt to Linnaeus is apparent in the *Origin of the Species*, where he describes the economy or *polity* of nature. Two hundred years later the definition seems straightforward: ecology is the study of the interactions between organisms and their environment. More specifically, it is the scientific study of ecosystems: networks or webs of relationships among organisms and forces at a wide range of scales and high degree of complexity. A basic premise of this notion is that nature acts systematically, hence the term "ecosystems". Yet not all 18th and 19th century thinkers adopted this premise uncritically. The philosopher Immanuel Kant understood the system's premise to be nothing more than a working assumption. He wrote, "In respect to her empirical laws nature has observed a certain economy that to our judgment appears fitting... and this presupposition... must precede all comparison. The judgment [Kant continued] orders appearances according to the universal, though at the same time indeterminate principle of a purposive arrangement of nature in a system, as if nature were something favourable to our judgment... [he then admitted that] without some such presupposition we cannot hope to find our way." Albeit necessary for the progress of modern science, the presupposition of nature as a *system* congenial to reason is still a conjecture, one not assumed by earlier scientists and questioned by some contemporary thinkers. A second premise of the modern view of ecology is the radical dissimilarity between the human and natural worlds. That, too, is a notion that needs to be questioned. Arguments for cultural ecology posit not only correspondences between these two spheres, but their shared basic premises.

**Oikos + Logos**

While the word ecology is no more than two hundred years old, the roots from which the modern compound originates date from two thousand years

Andrew Wyeth, McVey's Barn, 1948.


ago. Ecology (first used in German) derives from two ancient Greek terms, *logos*, a term I will not address, and *oikos*, which signified nothing so grand as the ecosphere or totality of natural systems; actually, just the household. However, that was not exactly nothing, still less a mere house – instead, oikos was one of the centres of human existence, comparable to the *agora*, yet not the citizen's political, but domestic centre. Eco as oikos + ology as *logos* would have meant something like "discourse on home life", had the term been used. Two millennia ago, economics, also a cognate, did not involve the management of finances (or not that alone), but the knowledge and practice of domestic stewardship. Xenophon's famous treatise *Oeconomicus* presents a full account of a well-run house, farm or estate. It is a great book for architects, even today's. Written at the time of Plato's dialogues, it is basically a treatise on estate management, ranging over topics such as the arrangement and storage of furnishings and supplies, the hiring of slaves, the cultivation of soils and the management of one's family. *Oeconomicus* was always important to architects because its topics included concerns that are central to the discipline: the place, organisation and orientation of buildings. Understanding these issues meant taking account of the environmental and extra-environmental characteristics of the building's location, its climate (the character and calendar of seasons) and the site's history. Home economics for Xenophon also pertained to the ways a family cooperated with others nearby, for the life of one family or farm depended on interactions with neighbours, in the polis or outlying lands, even if the dealings were conflictual. Health was key, as was simple and teachable work. This list shows the cultural dimension of ancient estate management. One of the limitations of the book is that it did not address the city or public realm, which for the Greeks, no less than us, represented the most articulate or communicative embodiments of culture.

Among all the issues addressed in Xenophon's study, orientation is perhaps the key, insofar as it indicates that the source of order within the oikos was partly external to the house itself. This is obvious as soon as one considers the formative role of conditions of climate, neighbourly relations, language instruction, history and so on. David Orr, a contemporary ecologist, has said the world is not only a resource, but also the foundation for the order of every single part – whether one is thinking architecturally or socially. It was only in the 19th century that monetary matters began to dominate economic discussions, and the modern science of managing money came into being. Despite this, even today ecology means participating in the common good of the whole. Participation of this sort had an important name in ancient Greek discourse, *ethos*.

### Ethos

Ethos: is it the same as ethics? Yes and no. Yes, if the accent is on conduct and the decisions it involves and no, if we think of a code of conduct that should guide behaviour. We tend to see ethics as a set of rules; well-defined, measurable and capable of ensuring the common good. In point of fact, ancient writings on *ethos* did address the matter of cultural norms, but rather marginally. More important was the tension between habits and deliberation. The key point about such decisions was that there were no pre-established steps or rules that could reduce their difficulty or lessen the individual's involvement in the choice. I stress this point because we have just the opposite condition in the ethics of contemporary ecological design.

In architecture today one such code has been drawn up by the United States Green Building Council, called the LEED certification system. By definition providing third-party verification that a building or community was designed and built using strategies aimed at improving performance across all the metrics that matter most: energy savings, water efficiency, $CO_2$ emissions reduction, improved indoor environmental quality, and stewardship of resources and sensitivity to their impacts". Obviously, this is rather distant from the earlier sense of decision making in a context of shared understanding of the public good. The kind of *conduct* signified by the word *ethos* was particular and local, also less objectified than LEED criteria, which is to say personal, like diet. When cultural ecology is the topic of study, comparisons with health are often useful. Despite their superior science, doctors still ask patients how they feel. Treatment works when the individual feels right. Following standard procedures may or may not lead to this result.

Ethos was a kind of behaviour that was constituted culturally. A good translation of ethos would be "habit of dwelling". In our time, as in antiquity, habits are not only those actions that are performed repetitively; more important is their suitability to given circumstances. The important point is this: these circumstances and situations are both environmental and cultural.

Let me recapitulate my opening points: 1) that shared culture – not the command of analytical methods and design techniques alone – enters into decision making in design, even so-called sustainable design, 2) that it is senseless to embark on project making in our time without considering the design's ecological meaning, 3) that our current sense of ecology is rather recent and has its basis in the modern natural sciences, 4) that the ancient concept that is abbreviated by the modern pertained to not only environmental but historical and cultural conditions, signified by the fact that *oikos* meant household or estate and 5) that the physical premises of the *oikos* were always understood in connection with the environmental and practical affairs they accommodated and represented.

Because my goal is to circle back from remote antiquity to present times, I will now consider a more recent example, a small guesthouse built by students south of Stockholm after a design by the Norwegian architect Sverre Fehn. I

Philadelphia Street, 2005. © David Leatherbarrow

Sverre Fehn, Eco House, 1992, view. © Lars Hallén

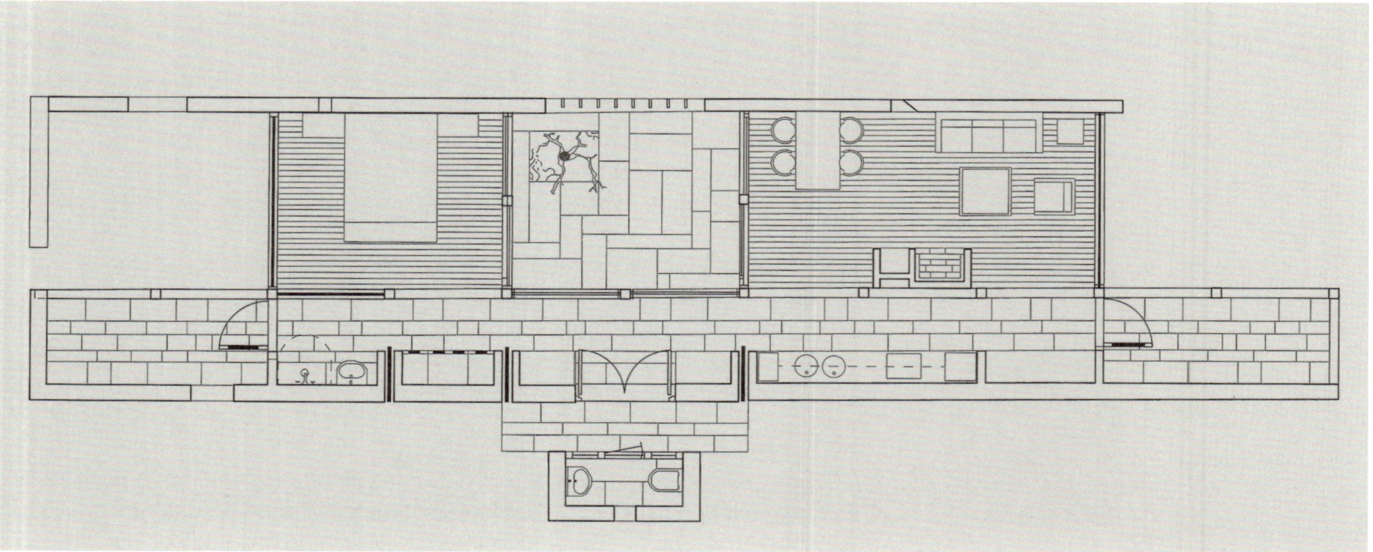

will consider this small work as an example of *ecology as a mode* of being in the world – both a place and a pattern, or web and way of life that shows the architectural potential of cultural ecology.

### Sverre Fehn's Eco House

Fehn's experimental holiday house in Mauritzberg dates from 1992 and was intended to be a prototype for a much larger development on the same site, a vacation and conference centre in the vicinity of an old castle. The design won a competition, but was not built as intended. Well after the initial project lost momentum, students from Scandinavian universities built a single prototype. The construction was completed in eight weeks. As the project stands on its own, it is a little hard to see its ecological order. Fehn himself observed that single solutions can hardly be said to be ecological if unrelated to wider circumstances. Still, I would like to use this project to describe Fehn's ecological approach. He is already known as an architect who was deeply concerned with architecture's cultural meanings. Maybe in this project we can see how the two correspond.

A number of critics have described the house's plan as "introverted". Two patios bracket glazed living spaces, all four align with a circulation corridor that runs fourteen meters in length and contains building services and the entry. Critics also seem to agree that the plan is highly rationalised. It is certainly economical (in the "spend less" sense of the word): only 53 $m^2$ for a two-person dwelling. Despite the economising, the bedrooms have a remarkable generosity of space, resulting from their openness on two sides. All the spaces are made to work very tightly and efficiently, the corridor in particular doubling as part of the kitchen space, bathroom and bedroom, when the sliding doors are closed. The services are organised along one wall, without specific, delineated rooms: a standard niche depth is deployed for kitchen appliances, storage and the bathroom units (the toilet in this version was situated just outside).

The compactness of the plan introduces one of the commonplaces of ecological design: avoiding waste. Yet, in this case, the plan's several elements – the linear spine, partitions and doors (not walls), enclosed courts, and corridor doubling as several rooms – also have cultural purposes. Fehn's aim, Per Olaf Fjeld writes, was to have "glow of hospitality kept inward".[6] This term invokes a Scandinavian sense of living in northern climates. It is also plain that the courtyards allow the rooms to breathe, while maintaining privacy.

Sverre Fehn, Eco House, Competition entry, 1992, plan. © National Museum of Art, Architecture, and Design, Oslo

Sverre Fehn, Eco House, 1992, plan. © Per Olaf Fjeld

Can the building's tectonic order be described as ecological? It is a simple and traditional mode of construction in which the spatial module is identical to the construction module; each room is two or three bays, which is to say either four or six meters. Even though its long walls have a solidity and thickness that give the impression (and thermal benefits) of load-bearing construction it is essentially a frame building, by virtue of the modularity and regularity of the column grid, as well as the truss system for the vaults. The first version of the design specified brick within the intervals of the frame, but when built the front and back walls were made out of clay. I will return to the matter of materials shortly.

In a small vacation house of this type it would hardly seem necessary to have any mechanical system for managing the internal environmental quality. And, in point of fact, the building is naturally ventilated and cooled. In addition to providing heat on cold nights, the fireplace provides the house with a social centre, in between the settings for dining and sitting together. One can see in the plan and views that the partition walls that run along the short sides of the bedrooms have operable windows and doors. Because of the plan's relatively shallow depth, these openings are able to keep the building's air and humidity at comfortable levels. Rolling wooden blinds on the windows modulate the light

– particularly when glare might be a problem – and they allow for privacy. The absence of glazing bars on the outside window walls increases the sense of openness: in the bedroom of course, which looks into an open court, and more widely in the living room, which opens toward the wide landscape.

Now I would like to return to the question of materials. Here, too, one can recall commonplaces of ecological thinking: the use of local materials for building construction. Obviously this is not the case with the glazing. Nor is the structural timber specific to the region, even if it is typical in Scandinavia. The materials of the flooring, roofs and infill walling are, however, site specific. Local stone is used on the floors, larger dimensions and an irregular pattern in the court, smaller sizes and parallel lines along the corridor. The roof is interestingly composite: cork, bark and earth. In another version of this house, also built by students, but in the city of Kolding, the roof was made from lathing, tar paper and lapped roof boards. Like the roofs, the walls are aggregate: 10 percent clay and 90 percent straw, covered in plaster, inside and out. Although the work was undertaken by students, Fehn supervised the construction and personally demonstrated the plastering technique, having learned the art decades earlier when he studied Moroccan vernacular architecture.

Earlier I observed that Fehn himself said that one cannot understand or design a house or building ecologically without taking account of its wider frame of reference. In some instances the architect can partly determine that wider frame of reference. The house I have been describing was envisaged as part of a very large ensemble. In the entire complex there were to be 250 houses. Obviously houses were not the only components of the vacation centre; it was to have a golf course, tennis courts and equestrian facilities, as well as places for relaxation, entertainment and retreat. The location near a forest at the edge of Bråviken Bay provided a context for these sorts of activities. Basically, the layout was planned as long rows of houses, conforming to the fall of existing terrain, with the central spine of each house or pair more or less perpendicular to the access drives or contour lines of the slope. Yet despite its topographical specificity, the project presents rather insistent repetition, as if abstract principles were no less important than concrete conditions. I realize this is a controversial assertion, when both ecology and tectonics are considered. Yet I believe that engagement with the location was attempted thorough compositional means that also allowed for the relative autonomy of the individual parts. The siting strategy is this: not anywhere, not only there. This type of involvement *and* separation is comparable to what we have seen in the plan and construction of the individual units: the plan type combines axial organisation with courtyards; the first is typical of Fehn's other Scandinavian projects, and the second is reminiscent of Mediterranean models. Similarly, the materials are alternately specific to the region and typical of modern European (even international) building. Localism is part of Fehn's sense of ecological design, but not all there is to it. Estrangement is there too.

### A Scandinavian House with a Moroccan Background

Although it is both necessary and natural for an architect to approach each project on its own terms, no one begins a design apart from a history of projects, an established manner of working, no matter how newly- or well-established that manner may be. What is more, this history includes one's own way of working and that of others, architects one has studied, been taught by or collaborated with. This background – this disciplinary context or institution – always qualifies the work's uniqueness, making it specific and general, of its moment and part of a tradition. I have already alluded to some of the background for Fehn's design: his knowledge of plastering technique, for example, was gained in his youth, during an important visit to Morocco.

Sverre Fehn, Eco House, 1992, interior. © Lars Hallén

Sverre Fehn, Eco House, 1992, interior. © Lars Hallén

Why North Africa? He once explained that he found his way there via Leo Frobenius, the author of several important ethnographic studies, well known to figures as different as Mies van der Rohe and Aldo van Eyck. He made the trip in 1951, after receiving a state grant. Some sense of the importance of this trip can be gathered from a four-page text he wrote, the longest piece of writing he ever published, called "Primitive Moroccan Architecture". He also lectured about what he learned there and returned in 1987 with a group of students from the architecture school in Oslo (AHO, Arkitektur- og designhogskølen i Oslo). In his essay he explained that he first went to Morocco "not to discover new things but to recollect what had been forgotten". He recalled that "when you enter the valley at sunset and hear a man call from the highest rooftop, you will still think: I don't know anything about this". Still, some sense can be made of it, for "the only answer to this architectural simplicity and clarity is that it exists in a culture and seems timeless".[7] I believe this early realisation of architecture in and out of its local culture, which is to say, culturally specific and natural, guided his work for decades to come.

But there was also a more specific insight gained in Morocco that pertains to the ecological house of forty years later. After a return visit in 1987 he wrote: "There is a zone between the [Moroccan] plain's earthen walls and the mountains. Wood construction slips through in this belt..." That sounds a lot like the house we have studied. About the construction of the typical Moroccan house he observed: "This is a regional use of materials, the houses literally speaking growing out of the ground they are standing on... The colours of the village are the same as the soil."[8] Not only do the walls support the roof, they also break the sun and provide a support for your back when you sit on the floor, just as they support racks for drying fruit and the drawings of young children. Meals, though, are centred around the table and carpet, elements of the house that are less permanent than the walls, but no less important, culturally speaking. Here tectonic, ecological and cultural concerns interweave and co-determine one another.

As if he were anticipating work that would come four decades later, Fehn wrote in his essay on Moroccan houses that primitive architecture could be likened to modern architecture, despite the fact that primitive culture is beyond the grasp of "we moderns". How can the two be understood together? The common ground is that works of both periods are specific and non-specific to their culture. The non-specificity becomes apparent when you consider its engagement with the natural world, something architectures of all ages share.

### The Influence of Prouvé and Le Corbusier

The problem of modernity – particularly the possibilities of modern building technology – had been on Fehn's mind in the years before the Moroccan trip. After graduating he spent a year working in Paris. His former teacher arranged for him to obtain a position in the studio of the great engineer/architect Jean Prouvé. This may seem an odd connection, given Prouvé's involvement with industry, specifically the aircraft industry, but Fehn was attracted to the office because he himself had developed a strong interest in prefabrication.

Prouvé's focus on economies of construction can be seen to have had an influence of Fehn's thinking, as did a number of more specific aspects of design: façades with adjustable elements (as in Prouvé's Mozart Square building, developed while Fehn was in the office), his tendency to develop highly compact plans, particularly in standardised housing, and his invention of highly rationalised service cores. All of this can be seen in Fehn's later works: the compact plans of the Norrköping and Schreiner houses, for example, also their service cores, and the adjustable elements on all of his façades. Over the years of his practice he seems to have shifted from centralised to linear organisation

Sverre Fehn, Schreiner House, 1959-63, window vent detail. © David Leatherbarrow

Sverre Fehn, Eco House, 1992, view. © Lars Hallén, Nordic Museum, Stockholm

of services, along a circulation spine. Representative examples would be the Busk, the Bødtker and finally the Eco House plan. In each of these cases the distribution of elements is ruled by the principle of economy. But in these cases, as in the Moroccan precedent, saving is coupled with cultural practice; the compact configurations were always (also) interpretations of patterns of life judged to be well-balanced.

I have one last reference to the background of Fehn's way of working: Le Corbusier. While often overlooked in conventional histories, Le Corbusier had a deep concern for the building's engagement with the natural world. There are many ways of approaching this topic. Just now I want to mention a single element that attests to his lifelong concern with natural ventilation: the vertical slit window, with a screen and operable panel. He experimented with this device in several projects, most notably perhaps in the small cabin he used for a summer retreat. The element achieved its final form, however, in the Brazil Pavilion, from 1956. Strikingly, more or less the same detail can be found in Fehn's Schreiner House. Both architects were concerned to let the walls of their buildings breathe. It was one of several ways they were to participate in the play of natural forces. That participation – at once environmental and cultural – is what I take to be at issue in architecture we might today want to call sustainable.

**Design Principles for Cultural Ecology in Architecture**

I would like to close by offering a set of principles of design, based on these several examples of what cultural ecology can be in architecture:

- Design's task is to integrate a way of life into the web of life.
- Personal and public habits generate invention in design and are a measure of its success.
- Scarcity may suggest sacrifice, but it can also lead to sharing.
- Renunciation is no more important than expression, if both are ruled by a sense of proportion – another word for sharing or giving each person or place due measure.
- The selection, treatment and assembly of materials, together with the configuration and qualification of settings, can be evidences of ecological order, if cultural norms guide the orientations and involvements that define the design.[9]

**David Leatherbarrow**
is professor of architecture at the University of Pennsylvania School of Design, where he teaches courses in architectural theory and design studios in the graduate and undergraduate programs and directs the PhD program. David Leatherbarrow holds a PhD in Art from the University of Essex and is author of a series of books including *Architecture Oriented Otherwise* (2008); *Topographical Stories, Studies in Landscape and Architecture* (2004); *Uncommon Ground, Architecture, Technology, and Topography* (2000) and *On Weathering, The Life of Buildings in Time* (1993).

Le Corbusier, et al, Brazil Pavilion, 1956-59, window vent detail. © David Leatherbarrow

Le Corbusier, et al, Brazil Pavilion, 1956-59, window vent detail.© David Leatherbarrow

1 __ At the University of Pennsylvania I co-teach a course on cultural ecology with Richard Wesley. Our title is: Cultural Ecology: uncovering the roots of green building in the early modern movement. Two premises guide the arguments we set forth: 1) that the ecological concepts derived from research in the natural sciences need to be augmented by cultural understanding if ecology is to have a significant role in architecture, and 2) that these considerations are not new to architecture but can be seen in the works of early modern architects, despite the fact that this dimension of their work has been neglected in existing histories. A co-authored book called *Modern Architecture as Cultural Ecology*, based on this course, is forthcoming.

2 __ Julian Steward, *Theory of Culture Change* (Urbana: University of Illinois Press, 1955), p. 31.

3 __ Steward, *Theory of Culture Change, p.* 32.

4 __ Immanuel Kant, *Critique of the Power of Judgment* (1790), first introduction, ed. P. Guyer (Cambridge: Cambridge University Press, 2001), pp. 18-19.

5 __ https://new.usgbc.org/leed

6 __ Per Olaf Fjeld, *Sverre Fehn. The Pattern of Thoughts* (New York: Monacelli, 2009), p. 98.

7 __ Cited in Per Olaf Fjeld, *Sverre Fehn*, p. 42.

8 __ Per Olaf Fjeld, *Sverre Fehn*, p. 40.

9 __ These principles elaborate those offered in an article I have co-authored with Richard Wesley, “Frameworks of Performance and Delight,” *Harvard Design Magazine* (Cambridge, Mass.: Harvard University Press, 2009), pp. 84-95.

Karl Christiansen

# Nature, Culture, Tectonics of Architecture

The present article advocates the thesis of a close relationship between the movements of the Earth's crust and the creation of architecture, on both the conceptual and the concrete level.

*No matter how we twist and turn it, we build our houses in relation to the Earth's surface, immediately below, on, or above it.*

*When seen in the perspective of the dimensions of the globe, including the Earth's 12,750-km diameter, the distance we add to the surface of the Earth when we build is merely a trifle.*

*Ordinarily, it is indeed only the outermost layer of the crust in which we show any interest with regard to building architecture. Is the foundation soft or hard? What about topography?*

*But these aspects, form and condition result from the workings of the Earth in its deeper layers: the movements of tectonic plates in relation to one another. In what way is this related to tectonics in architecture?*

## Mythic

"χάος", said the ancient Greeks. "In the beginning there was chaos." Out of this chaos a ball shot out. This ball was Gaia: Mother Earth. And thus the world was created, even though time as yet stood still.

Chaos was not chaotic in our sense of the word. It was more like a chasm, a limitless opening, an infinite, empty and immeasurable space, in which what had been created appeared.

But how could the Greeks actually know this? Well, for very good reasons they could not. Nevertheless, this image is almost uncannily similar to what researchers discovered several thousand years later: an evolution progressing over so many incomprehensible billions of years that, in this period, to us, time is almost standing still.

## Geological

The origin of our globe was the explosive contraction of a primeval nebula. This happened 4.5 billion years ago. This contraction created a rotating disc, which developed, centred around the sun. The planets were created by the collision of matter; the accumulation of rocky material orbiting the sun. The same applies to the Earth, which was formed 10 million years after the contraction as a liquid glowing sphere in which the denser components had settled, we might say, at the bottom, and had formed the core, whereas lighter matter floated on top and formed the outer elements. Since then the red-hot, liquid mass has cooled, possibly due to an increased distance to the sun, but also due to a bombardment by meteorites of ice, which meanwhile created the oceans. This cooling process formed a hard crust: the Earth's crust. The Earth's crust is the uppermost part of the lithosphere, which also includes part of the subjacent mantle. The lithosphere is divided into tectonic plates, which slide on the softer asthenosphere below.

The division of the lithosphere into individual plates occurred when the hard crust cracked. This was an inevitable result of the cooling which caused the material of the Earth's crust to contract, while the glowing asthenosphere below retained its volume and resisted compression. Cracks in the crust were the only way to avoid an actual explosion.

### Tectonic plates

Tectonic plates constantly change their position in relation to one another. This is subject to the glowing core of the Earth, which is still in motion because the temperature of the mass is not homogenous. It is warmer near the centre than near the surface, and the poles are the coldest areas. The cooled substance sinks and the heated material rises. Thus everything is in a state of continual, dynamic movement.

In general, tectonic plates move very little, and for us, fortunately, this movement is indiscernible. However, it is not outside the rules that movements become so pronounced that they have consequences for us who inhabit the Earth. Locally, the result may unfortunately be disastrous.

The more violent of these movements in the tectonic plates may radically change the appearance of the Earth and, in particularly exposed areas, they may consequently change the conditions of habitation. This phenomenon simply generates a radically new form of landscape. Overall, this happens in three different ways:

- Compressional. Ad-
  Plates move towards each other. This typically creates mountain ranges. We can call this activity “ad-dition”: material is added and accumulates locally. This is the way the Andes Mountains were created. So were the Rocky Mountains a bit further to the north.

- Extensional. Sub-
  Plates move apart and cracks appear. We call this “sub-traction”. This process leads to the disappearance of material. This is exemplified by the clearly identifiable cracks at Tingvalla in Iceland.

- Transformational. De-
  Plates are displaced sideways in relation to one another. In this case “de-formation” is an appropriate term. The locally available material is rearranged and thus creates a new, changed form without material being subtracted or added locally.
  The San Andreas Fault near San Francisco is an example of this. In addition to this, there are “subordinate processes”, which are, however, all variants or hybrids of the above three. For instance, the Hawaiian Islands are the result of a so-called "hot spot": the liquid lava managed to burn through the very centre of the Pacific Plate. So much material was added to this

*"In architecture, the material is indeed also, at least to begin with, gathered in nature (wood, stone, iron, earth, etc.). However, by now, the activity has become artificial: the processing of the material (sawing, chopping, casting, heating, hammering, digging, etc.). The result is the architectural form: some 'other' geometry."*

location in the middle of the Pacific that parts of it are now above sea level. Where Iceland is now located, tectonic plates initially moved away from each other: material was subtracted. But, immediately after this, the result was a volcanic eruption. Material was added as magma from deeper-lying geological strata found an access point to surge upwards through a crack, thus creating Iceland.

**Material – Activity – Form**

Any production of specific physical form follows a specific pattern. *Material* is processed through *activity*, and *form* is created.

For the processes of geological tectonics, the material consists of the plates of the Earth's crust. The activity is the movement of tectonic plates in relation to one another: pressure, pull and displacement. The resulting form consists of mountains, cracks and faults.

**Nature – Culture**

The aforementioned are all natural phenomena: i.e., forms created through natural activity. These are, at first, in opposition to architectural activities, because architecture is not created as a result of natural activities, but rather as a result of cultural activities. Architecture is therefore the opposite: nature made art-istic or art-ified.

There is, nevertheless, a striking similarity between the ways in which the two activities produce concrete form.

The decisive differences between the two activities are, of course, the temporal and spatial, which are staggering. Whereas we are powerless when faced with the movements of tectonic plates, we can influence the movements of the concrete floor. Not whether it will shrink and crack and move (because it will), but in what way. On our temporal and spatial scale we have an opportunity to harness the inherent activity of materials for the purpose of creating an intentional form. We must take into consideration the latent virtuality of individual materials, regardless of whether we use them on their own or, what is more common, in combination with other materials. To accept the crack as a condition, but to control where it should appear, is inherent in the creative architectural activity of humans.

In architecture, the material is indeed also, at least to begin with, gathered in nature (wood, stone, iron, earth, etc.). However, by now, the activity has become artificial: the processing of the material (sawing, chopping, casting, heating, hammering, digging, etc.). The result is the architectural form: some "other" geometry.

But here the pattern is the same. You take a *material* and process it or re-shape it through an *activity,* and the result is a *form.* Furthermore, the way this activity creates form is either by ad-dition, sub-traction, de-formation and/or a mixture of these processes.

We daily experience the principle at the heart of the phenomenon which created tectonic plates: in all other, now local, surfaces, which also form cracks, and break if the area is too large compared to the thickness of the surface, the composition, differences in temperature, humidity, dryness or the condition of the base. Furthermore, all these surfaces are constantly moving uncontrollably in relation to one another. Most materials change when exposed to changes in thermal conditions. Many expand to some extent when exposed to increased humidity, and contract and shrink when moisture is removed. Many materials expand and become more plastic when heated, and vice versa when cooled. In brief, thermal conditions are conditions we need to take into consideration when we create architectural form. Specifically, "If you consider joining two different materials, you should dig the difference!"

**Material – Technique – Form**

In architecture, the activity we carry out in order to process a material is the technique, the technical processing. In this context, technique refers to human beings joining means of work, objects of work and manpower in the work process.

At the same time, technical processing is the stage which decisively separates culture from nature. Only then can we talk of architecture. Consequently, the forms created by technique are always significantly different from the forms found in nature. No matter how much we struggle and strive, it is always obvious when human intelligence and skills are behind the production of a specific physical form.

— Nature has its forms, just as culture has its forms.

The three parameters, material, technique and form, are common features. They form one combined set, which constitutes one of the smallest elements of architecture: an axiom.

It is possible to imagine a great many things in relation to concrete physical architecture, but it is impossible to imagine the absence of a material, a concrete substance. Furthermore, this concrete material presents itself as a concrete form. Meanwhile, for this form to be able to lay claim to the name of architecture, it must have been brought about by means of technical processing, or at least intentional handling. That is surely the way it is.

**From the inside**

However, not everything we produce is fundamentally tectonics.

— It is possible to imagine that the same form is produced in "another" way than that which belongs to it or its material.

— It is also possible to imagine the same form produced in another material than the one which belongs to the form and the way it is produced.

— Finally, it is possible to imagine that another form is produced in the same material as the one which belongs to it and the way its material has been produced.

But does this mean that material, the way of production (technique) and form are interrelated in one specific way? If it is tectonic, the answer is yes! As has been demonstrated, the parameters of material, technique and form are probably always present, but they can relate to each other in unequal ways. That is to say that one parameter may have been given a higher priority than another or all of the others, which, for their part, have to make "a nip here, a tuck there" to accommodate the first parameter.

This indicates an imbalance. For instance, the desire to obtain a specific form can be so stubborn that it "twists the arm" of the other parameters, and the result becomes incoherent. Metaphorically speaking, the three parameters can be seen as three straight lines reaching towards one another in an attempt to form an equilateral triangle. But the triangle leaks from one, two and often all three corners. The starting point, the desired form, is admittedly present, but without substance, as a postulate we would say, because the form stands alone, unsupported by any tectonic creation in the sense that it is one with its maker (the parameter(s) which have created it). It remains an expression of something it is not. It has become an image of something without actually being this something.

**Karl Christiansen**
is a Danish architect and researcher. Since 2004 he has been a professor at the Aarhus School of Architecture. Before devoting his time to research and teaching he practised architecture in Norway. Within a Nordic context he has been an important exponent of the development and discussion of tectonic thinking as the core element of architectural practice. This work has led to publications such as *ArkitekturKonstruktioner* (1994) and *Dodekathlos – om arkitekturens tektonik* (2004). Karl Christiansen is also well known for his experimental work developing new principles for casting concrete.

Material

Technique

Form

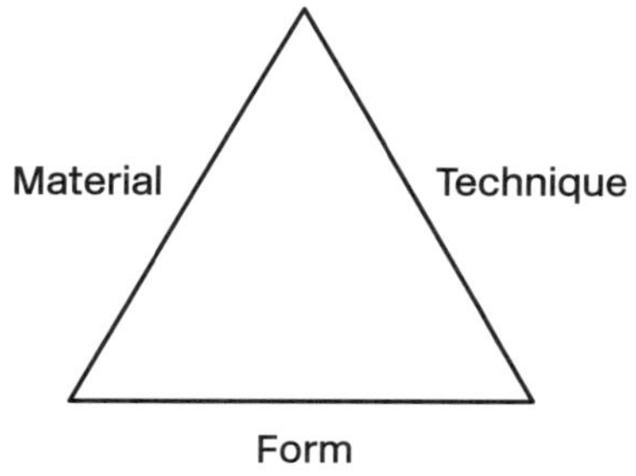

Christiansen, Karl. *Arkitektur – form og teknik*. Arkitekten. 1995 (17).

But isn't this alright? Well, of course it is! But it still is not tectonics. The tectonic form is not a detached image which has been attached to something, which this image does not itself express. The tectonic form comes from the "inside".

That is how geological tectonics work. Here the tectonic form is a direct result of the processing by the activity of the local material, no more, no less, but precisely this. The talus is an obvious example of this. It is the form generated at the foot of a mountain by fragments detached from the mountainside, which have gathered and created a subsidence form created by the total of circumstances: this, only this, and this is enough, because more is too much and less is not enough.

The form comes from the "inside". In this way a volcanic crater is a tectonic form, whereas a crater formed by a meteorite is not. Nonetheless, they both still occur as concrete forms on the surface of the Earth.

Tectonics is connected to architecture in a similar way. Etymologically, the concept of tectonics never primarily describes an aspect of form. Instead it always describes the circumstances that create the form, its premise: The craftsman, the artist who processes a material, the person who manufactures something, composes something, builds and creates something. Tectonics indicate the schemer and the poet, who both create on the basis of nothing, even though it is in particular the parameters that bring about the art of building as art, as architecture, which are in focus etymologically. Form is created on the basis of its premise(s). In this way, form is created from the inside out. Tectonics is a product of the genes. It is generic: a type of genetics based on the Greek act of creation (γένεσις) as opposed to giving birth (γέννα). This creation presents itself as an unhidden truth: that is to say, a truth which distances itself from correctness. This should be understood as follows: the latter, if it is hidden, can be verified by some third party or element. It is like a witness saying, "What he says is correct". The first-mentioned, however, only speaks for itself. The Greek called this truth αλήθεια (alithia). Alithia is when truth is happening, and in this way tectonics happens.

Claus Bech-Danielsen

# An Abstract Culture's Search for Concrete Roots

Weissenhof 1927: For modernist architects, hygiene requirements became an integrated part of architecture. For example, at the Weissenhof Exhibition in 1927, electricity was highlighted as a clean source of energy. © Claus Bech-Danielsen

## Modernism – Architectural Solutions to Concrete Environmental Challenges

Today architects accept the fact that construction and urban planning must incorporate sustainability. All major architecture firms have a strategy that focuses on sustainability, and virtually all architecture competitions include sustainability as a central criterion. One could say that sustainable development has become a premise on which all architecture is based.

Things have not always been this way. In the late 1980s, when I first began addressing the issue of environmental challenges in my work, I could not find many fellow architects with whom I could discuss my thoughts, and when in the early 1990s I dealt with the issue in my PhD dissertation, I was met with great scepticism from the architecture world. Sustainability was associated with urban ecological grassroots projects. Such projects were known for their "anti-aesthetics", and the reaction of the architecture community was to turn its back on anything that said "environmental". Instead, architects focused on developing the new-modernist architecture of the time, which they considered to represent a stark contrast to the environmental challenges of the time.

That neo-modern architecture represented a contrast to environmental efforts is a paradoxical viewpoint, as modernist architecture and planning originally grew out of a desire to combat environmental issues. The modernists' grip on architecture was based on environmental pollution, which was causing serious health problems in society at that time. At an urban level, health problems were caused by smoke, noise and wastewater from industry, which in the densely populated cities existed side by side with residential buildings.[1] Moreover, cities lacked green areas (ibid.) and were dealing with congestion problems from increasing traffic (ibid.). In dwellings, health problems were caused by poor access to daylight,[2] lack of ventilation,[3] damp from basements and smoke from kitchens and wood-burning stoves.[4] The modernists wanted to solve these environmental challenges and create cities and buildings which were healthy for both body and soul.

The environmental challenges to which the modernist architects reacted were tangibly prevalent in contemporary cities and dwellings. They could be seen, smelled and heard. They could be experienced by the senses. This enabled modernist architects to link environmental aspects to the development of concrete architectural qualities, which could also be experienced by the senses. In dwellings, this meant that new types of glass and large windows were used. This was partly in order to increase the amount of daylight in the building and thus improve the health of residents, and partly to take advantage of the experiential qualities of daylight, creating spatial relationships between the interior and the exterior. At city level, the modernists solved contemporary environmental problems by zoning the city, which also provided them with an opportunity to create residential areas with light and space, as well as easy access to scenic qualities, i.e., experiential qualities.

The issue of environmental sustainability now ranked high on the modernist agenda. Moreover, what we today refer to as social sustainability was also a central issue for the modernists. Modernist pioneers demonstrated strong political commitment; they wanted to create a non-hierarchical architecture, which could constitute the framework for a democratic society. Thus they strove to develop a concept for cheap mass production of housing in order to create good and healthy homes everybody could afford.

Thus architects have actually been addressing sustainability for almost 100 years.

Weissenhof 1927 © Claus Bech-Danielsen

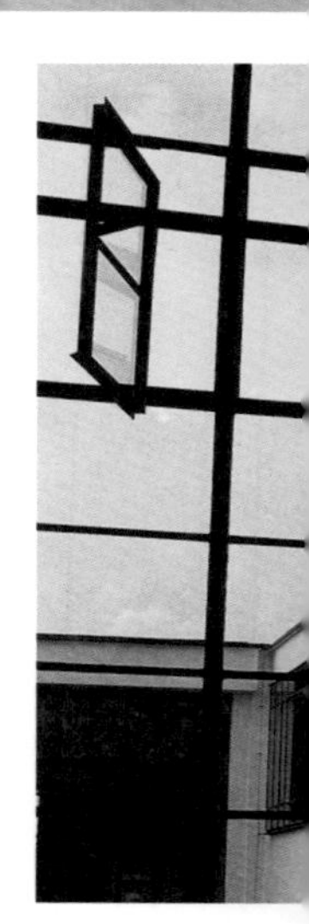

### Sustainability – a New Focus on Time and Space

When the welfare society's construction of housing took off in earnest in the late 1950s, this housing remained true to modernist architectural and urban-planning ideals. A great number of good, functional buildings had been constructed in the 1930s and '40s and, in line with increasing affluence, construction processes were to be further industrialised and buildings mass produced so as to meet the postwar demand for more housing.

In many ways, postwar construction was successful: for example, in terms of improving social and environmental sustainability. The severe housing shortage in Western societies immediately before and after WW2 was overcome, as were the environmental problems that characterised cities and dwellings in the wake of industrialisation. However, these environmental problems were not really solved; they were simply moved. With the zoning of cities, polluting industries were moved away from residential areas, and through refuse collection systems, sewage systems and tall smokestacks, waste and other environmental pollution were conducted away from cities and into the surrounding environment.[5] Later in the 20th century, not only was pollution "exported" from our immediate surroundings; the sources of pollution were also exported by outsourcing industrial production to other countries. In fact, pollution was still a problem. It had merely been moved far away from our cities to very distant locations.

However, it soon became clear that environmental problems could not be solved by building tall smokestacks, by exporting pollutants to distant places, or by moving production farther away. Even the most distant rainforest, the deepest sea and the most remote outer atmosphere were affected by environmental pollutants, and we were forced to come up with new solutions to this challenge. Environmental pollution had become a truly global problem and required action at international level.

As a result, in 1983 the UN set up the Brundtland Commission, and four years later the commission published its report: *Our Common Future*.[6] This report highlighted the need for a sustainable transition, and *sustainability* was defined as the way to meet the needs of the present and the environment, without compromising the ability of future generations to meet their own needs (ibid.). Sustainable development basically concerns developing a new time perspective, which incorporates the needs of tomorrow in the actions of today.

The Brundtland Report put environmental issues on the political agenda of Western societies and, because construction is a major contributor to environmental impact issues, it was essential to include this area on the agenda as well.[7] The requirement put forward in the Brundtland Report to include the needs of the future in the actions of today led, for example, to the development of overall economic models that add the future operating costs of a building to the construction price. This new time perspective also led to the development of new theories of design and architecture, including McDonough and Braungart's "Cradle to Cradle" approach.[8] This approach introduces taking a cyclical perspective in construction and focuses on the overall lifetime of a building: from procurement and processing of building materials to the actual construction, use and demolition of a building, and finally subsequent reuse of materials.[9]

In modernist architecture, environmental improvements went hand in hand with developments in architectural qualities. For example, the impressive windows of the Bauhaus School, which provided a healthy indoor climate (an abundance of daylight and ventilation), let in the daylight and led to new spatial linking between outdoors and indoors. © Claus Bech-Danielsen

This new perspective on time has been further developed in life-cycle assessment (LCA) analysis, in which the overall environmental impact of materials "from cradle to grave" is mapped. This perspective has also been further implemented by developing tools and computer programmes that allow architects and other relevant players to focus on the environmental impact of their architectural decisions.

The tools mentioned here assess the overall environmental impact of construction, including how procurement of materials affects the local environment at the extraction and processing site. For example, when using aluminium, the fact that the extraction of bauxite ore leads to significant landscape destruction is taken into account.[10] Moreover, the subsequent processing of bauxite to aluminium is very energy-intensive, entailing major implications for the greenhouse effect. This aspect is also given consideration (ibid.). This means that the architect cannot focus solely on creating qualities in a specific building at a specific site. The architect must increasingly incorporate the impacts of the construction process on other places, which may be located far away from the actual construction site.

Thus, sustainable development has not only led to a changed perspective on time; it has also led to a changed relationship to place and space. Focus on the concrete building site, the concrete place, is extended to include *many* places: in abstract space.[11] The requirement for incorporating sustainable development in the design process forces the architect to include a distant future in current actions as well as in a global context. *Sustainable development in architecture leads to a new approach to both time and space.*

### Abstract Environmental Problems Today

The new perspective on time and space renders the architect's work more complex. Moreover, the requirement for the architect to focus on issues that are related to abstract space may entail a risk that the focus is shifted away from the development of basic architectural qualities ultimately to be experienced directly in a concrete place and in a concrete building.

This risk is increased by the nature of new environmental challenges. The environmental challenges of the industrial society were tangible; they were visible and present in cities and local neighbourhoods in the form of smoke, noise, odours and dust, which could all be experienced by the senses. In contrast, many of today's environmental challenges are more intangible. They include, for example, the greenhouse effect and the growing hole in the ozone layer, which both exist in a distant atmosphere. Other current environmental problems include pollution from radon radiation, endocrine disruptors and microparticles, which may in fact be found in our living rooms, but which cannot be seen, heard or smelled. Today's environmental challenges are increasingly imperceptible, and can only be detected when using sophisticated measuring equipment. In short, they are intangible and abstract.[12]

Moreover, the resource consumption of buildings has also become more imperceptible. Heating is supplied from remote power plants, energy flows through hidden piping and wiring under floors, and heating systems are controlled by technology located in dark basements. This means that the energy consumption of buildings is not perceived in the same way as earlier, when the flames from the fireplace could be seen, smoke from the chimney could be smelled, and firewood had to be chopped and stacked. We can no longer simply rely on our senses to understand the cycle of a household.

Due to the abstract nature of today's environmental challenges, which increasingly entail shifting focus to places far away from the construction itself, and due to the near-invisibility of resource streams, integrating solutions to today's environmental challenges into the architectural design itself poses a challenge. Architecture is a tangible art form which is experienced by the senses. It has to do with rhythm and textural qualities, and concerns the relationship between light and shadow, lightness and heaviness, soft and hard. It is, therefore, not immediately apparent how concrete architectural qualities can be related to solving today's abstract environmental challenges.

Environmental Efforts in Abstract Space

This is seen in many environmentally friendly buildings. An example of such a building is the "passive house", a building form that has received a great deal of interest in recent years. The concept of the passive house is to reduce energy consumption in buildings by increasing the density of construction, managing air exchange and optimising insulation. The concept of the passive house can be seen as a dogma that is to be applied to all passive houses. Thus this concept is in fact universal. In these buildings, focus is not on the specific qualities of a specific place. Instead rules are established to apply to all future passive houses everywhere. Again we can state that we are dealing with an *abstract space* rather than with a *concrete place*.

Moreover, the concept of the passive house is not based on the lifestyle and the behaviour of the users of the building. Instead, the residents themselves and their behaviour are subject to the building's technical solutions. For example, technical solutions control ventilation and, if the residents overrule the system by opening a window, the building's overall environmental performance will flounder. Thus, this green solution is first and foremost a response to ambitious goals set for the energy area, and only secondarily a response to concrete everyday needs.

Therefore, passive houses fall into the tradition of environmental solutions which were developed as a response to the energy crisis in 1973. Since then, energy savings have been given very high priority, and politicians have striven to reduce energy consumption in the area of construction by tightening building regulations. In the Nordic countries, efforts have primarily aimed at reducing heat loss from buildings, and requirements regarding insulation of buildings have been repeatedly tightened. By introducing a regulatory framework, politicians and civil servants have attempted to improve environmental results in many areas, in what can be described as a universal environmental effort that is being developed in an abstract space rather than in a concrete place.

### The Abstract Development of Architecture

What has happened is that environmental challenges have become increasingly abstract, and correspondingly environmental responses in construction are being based on abstract and universal goals. This is a general

The technical approach to environmental efforts was not developed with a view to solving the environmental challenges of one specific location. The starting point could be, for example, a wish to minimise waste heat from ventilated air. A heat exchanger is developed in response to this, which in principle can be used everywhere –i.e., a universal solution. The photovoltaics in the photo represent the same kind of approach. © Claus Bech-Danielsen

problem, which is not restricted to environmentally friendly buildings. This abstract and universal approach has characterised architecture since WW2, and the results of this approach have been vehemently criticised for several decades now.

Functionality in abstract space. According to Habermas, postwar, assembly-line housing focused on the functional in an abstract system, e.g., economic functionality, whereas modernism's original ideals were based on a desire to replicate sensory qualities, and the actual and concrete experiences of everyday life were forgotten. Correspondingly, in the construction of many sustainable buildings, primary focus has been on developing sustainable solutions in an abstract world (energy consumption and substance cycle), and the sensory qualities of the architecture have taken second place. © Claus Bech-Danielsen

This criticism was first voiced in the 1960s by the Postmodernist movement led by Robert Venturi, who identified a number of downsides in the assembly-line approach to buildings. The Postmodernists believed that the modernists' vision to rationalise construction processes had led to monotony, and that the original dream of simplifying architecture was now nothing more than an elitist debate, far removed from the modernists' original intention to create housing for the masses.[13] At the same time, phenomenologists such as Christian Norberg-Schulz[14] criticised the concept of standardisation and the fact that focusing on universal challenges led to neglect of the special qualities of specific places. All places were treated in the same way and thus we were losing the identity of place. According to Norberg-Schulz, this sense of identity is essential for our creation of a personal identity.[15]

Sociologists also voiced similar criticism. They believed that the assembly-line approach to buildings was based on ideas and abstract ideals that were far removed from the reality of everyday life. Jürgen Habermas[16] addressed this issue in his identification of the distinction between a "lifeworld" and a system. Modernists placed great emphasis on developing functional solutions, but there is a significant difference between what is functional in a concrete lifeworld and in an abstract system. In a lifeworld, "functional" means that objects meet a certain need in everyday activity, whereas in a system functionality is about meeting abstract goals: for example, economic functionality. According to Habermas, postwar, assembly-line buildings focused on the functional in an abstract system. This meant that architects and others lost sight of the original intentions of the modernists: to replicate sensory qualities and provide the individual with concrete experiences equal in value to the logic of the building (ibid.). As described, similar criticism was directed at environmental initiatives of a more abstract nature within construction: *The focus has primarily been on setting goals in an abstract system, where sustainability means, for example, optimising energy supply and the cycle of substances. The development of sensory qualities and aesthetic experiences has been secondary.*

At first, criticism of economic rationalism led to a political countermovement. This was in the 1970s, when architecture became political. Architects were more interested in discussing politics than in developing sensory experiences and tangible qualities in concrete buildings. Since then a number of attempts have been made to reinvent the fundamental qualities of architecture as well as the role of the architect as an aesthetic designer. This was certainly the case with the wave of Neomodernism, which characterised architecture in the 1990s. As mentioned at the beginning of this paper, the social and environmental commitment, so characteristic of modernism in its early day, was forgotten in the design experiments with white façades that were so prominent in the 1990s. Neomodernism became a picture *of* modernism, it was superficial. It was Postmodernism in a new guise.[17 18]

The wave of pragmatism, which surfaced in the early 21st century, led by Rem Koolhaas among others, can also be seen as an attempt to stifle idealised conceptions and abstract thought constructs. This movement is based on the goal to recognise "dirty realism". Instead of forcing an idyllic image upon our cities, we must open our eyes to the reality that surrounds us.[19] However, Koolhaas' approach to architecture is no less elitist than that of his predecessors. In his scholarly enthusiasm for dynamic and globalised reality, he forgets that not everyone belongs to the mobile elite,[20] and that realities are perceived differently, and vary drastically, from location to location.

Thus it would seem that it is hard to bypass the universal way of thinking and the abstract ideas we have developed. To succeed we must briefly disregard political ideologies and ideas about society. We must focus on architecture itself, on the very foundation of this discipline, and view with new eyes the creation of architectural quality. As I will discuss below, tectonics may play an important role here.

**The "New Beginning" of Modernism**

First, let me remind you that this was first attempted almost 100 years ago by the artists and architects who instigated the Modernist movement. They were searching for the seeds of a new civilisation by reaching behind existing thoughts and ideas. In this way, they wanted to go back to the original starting point, on which they could base a new architecture and a new civilisation. For example, this was a factor when Piet Mondrian restricted his paintings to vertical and horizontal lines and the three primary colours. He was searching for basic forms, which he believed were devoid of cultural preconceptions. In a similar manner, architects contemporary with Mondrian were also in search of a new basic foundation. The Swiss architectural historian Sigfried Giedion thus stated:

*"Modern architecture had to (...) find a new beginning. It had to re-conquer the most original elements as if nothing had come before"*.[21]

Or, in the words of Le Corbusier,

*"Our diagnosis is that, to begin at the beginning..."*[22]

As described, this led to new abstractions. Considering that the focus of this article is sustainability, it is worth noting that an important reason for such abstractions is to be found in the modernists' perception of nature. There is a clear link between Le Corbusier's aesthetic ideals and those of Paul Cézanne, who wanted to experience nature as a cylinder, a sphere and a cone.[23] Whereas artists from the previous period had focused on the outer appearance and form of nature, modernists sought a new starting point by looking at the principles that guided the inner structures of nature.[24] They shifted their focus from nature in their visible surroundings, the outer nature, to a more "inner" nature. For example, in 1913, the art critic Apollinaire wrote about the Cubists that

"*[e]ven though these young painters are still observing nature, they are not trying to copy it, and they are taking great pains not to replicate natural scenes...*". Rather, they were seeking "*... the traces of the non-human that cannot be found anywhere in outer nature.*[25]

Seeking an outer nature, which had been perceived for centuries as God's creation, no longer made sense.[26] The modernists were therefore seeking another form of nature, which a new culture could "cultivate". Let us dwell on the words "nature" and "culture". The word "nature" is derived from the Latin *nascor* and means "to be born, to come into this world, to arise, originate, appear." The word "culture" is also derived from Latin, from the word *colo*, which means "to cultivate and care for the soil and surroundings". Nature is the original element in which culture establishes itself via mankind's cultivation of it.

Looking at it in this light, it becomes evident that all cultures build on a foundation in nature, and therefore no culture can be stronger than its own natural foundation. If a culture loses its natural foundation, it no longer has anything to cultivate, and culture loses its meaning and falls apart. Clive Ponting describes this and lists a number of examples in his book *A Green History of the World*.[27] Ponting's point is that if a culture loses sight of its natural foundation and falls apart, a new culture cannot arise until a new form of nature has been identified for cultivation. A new take on nature is needed.

This is precisely what happened at the beginning of the 20th century. The modernists rejected the established perspective on nature and its focus on the nature surrounding us, the outer nature, shifting their focus to a more inner nature. This was expressed in the abstract art of the period. The "inner nature" in nonfigurative art was, however, so abstract that only very few people could actually identify it. Not many people could navigate at a level of abstraction at which nature was to be experienced as cylinders, spheres and cones.[28] At the same time, the emergence of a number of serious environmental problems reminded us that it is not viable to think abstractly about the very concrete nature of our surroundings. We were forced to acknowledge the concreteness of nature.

Over the past decades a new perspective on nature has emerged. This new perspective has pushed forward a number of art forms such as "nature art" and "land art". Here, art is expressed in a concrete reality, using nature's own materials and processes. A new perspective on nature comes to the fore, rejecting a world picture in which culture views itself as being in opposition to nature. Today's architecture makes a similar break with the traditional understanding of architecture as a contrast to the surrounding landscape. The clear border between building and surroundings, between city and the surrounding area, between culture and nature, is being erased. A new culture is emerging, and this culture is searching for its roots in nature.

### Tectonics – A Search for Architectural Roots

Today's tectonic approach to architecture also reflects this search for roots. Tectonics strives to identify a basic starting point for architectural creation. This requires us to focus on the basic building blocks of architecture: building materials and building units, as well as the structural engineering with which they are assembled. Moreover, it requires us to explore the possibility of creating basic architectural qualities and aesthetic experiences in the actual design of the building materials.

In contrast to the modernists, who were in search of a new beginning for architecture by abstracting from a number of concrete conditions,[29] today's interest in tectonics is seen as the quest for a new architectural starting point, which focuses on the basic components of architecture. Whereas the early modernists sought to investigate the established perceptions of the world, tectonics seeks to investigate the imposing steel surfaces to find the underlying structures and constructions. Seen in this light, tectonics can be seen as a "new new beginning", which stems not from wiping the slate clean and creating a mental *tabula rasa*, but from an inquisitive examination of the concrete discipline on which architecture was originally based.

Seen in this context, tectonics takes us, not surprisingly, back to antiquity. For example, the Greek concept of *techne* becomes central, emphasising antiquity's understanding of architectural creation, in which there is no distinction between method and objective.[30] Drawing on *techne*, tectonics focuses on how the design of building materials and structures in itself is an artistic goal. No distinction is made between artistic and building-technology issues or, in the words of Marco Frascari, "between construction and construing".[31] They are regarded as one, as a unity.

It is in this very attempt to reunite a divided architectural world, where there is distance between design and technology, that we find tectonics. The tectonic perspective strives to re-establish a united architectural approach. It is here that we find a parallel between tectonics and sustainable development, which also takes a holistic approach to development. Sustainable development crosses boundaries (for example, national boundaries and sectorial boundaries) and focuses on the interplay between cities and their surroundings. In short, sustainable development strives to link culture and nature.

Other proponents of the tectonics movement look even further into the past for inspiration. In another article in this anthology, Karl Christiansen examines the etymological origin of tectonics. This leads him to examine movements within geology, and Christiansen subsequently describes architectural creation as the replication of these movements.[32] This leads to yet another parallel between tectonics and sustainability. The word "geology" stems from the Greek goddess *Gea*. *Gea* was the primal Greek goddess who personified the Earth, the Greek version of "Mother Nature". It was this same Greek goddess whose name was given to the Gaia theory developed by James Lovelock in the 1970s.[33] This theory views the Earth as a living organism and proposes that all organisms and their inorganic surroundings on Earth are closely integrated to form one single and self-regulating system, thereby maintaining the conditions for life on Earth (ibid.).

In geology's examination of tectonics, the Earth is also seen as a living organism. Obviously, both theoretical positions lead to procedural and open views on architecture. In the perspective presented here, we find the starting point for a "new new beginning", which links sustainable development to tectonic qualities in architecture. It is about developing an architectural approach rooted in concrete and sensory conditions, and about recreating a building culture that interacts with its foundation in nature. Ultimately, we, as cultural beings, must once again find our roots in nature.

**Claus Bech-Danielsen**
Since 2008, Claus Bech-Danielsen has been a professor at the Danish Building Research Institute, Aalborg University. His research has focused on the modernist tradition of architecture and how this understanding of the world can accommodate an ecological way of thinking. He is the head of the Danish Centre of Housing Research, and holds a PhD from the Royal Danish Academy of Fine Arts, School of Architecture. He is also the author of a large number of books and articles. The work presented in this text is very closely linked to his book *Ecological Reflections in Architecture – Architectural Design of the Place, the Space and the Interface,* which was published in 2005.

1 __ Le Corbusier, *La Charte d'Athènes – Constatations du IVe Congrès* (Paris: Plon, 1943).

2 __ Christer Bodén, *Modern arkitektur. Funktionalismens uppgång och fall* (Helsinki: ArchiLibris, 1989).

3 __ Le Corbusier, *Towards a New Architecture* (Oxford: Architectural Press, 1989. Orig.: *Vers une architecture*, 1923).

4 __ Claus Bech-Danielsen, "The Kitchen: An Architectural Mirror of Everyday Life and Societal Development" (*Journal of Civil Engineering and Architecture*, vol. 6, no. 4, 2012), pp. 457-469.

5 __ The German specialist in ecological urban planning, Ekhart Hahn, describes this type of environmental approach as "The tall smokestack principle". See: Ekhart Hahn, *Ecological Urban Restructuring* (Berlin: Wissenschaftszentrum Berlin für Sozialforschung, 1991).

6 __ Brundtland Commission, *Our Common Future* (Oxford: Oxford University Press, 1987).

7 __ For example, in the energy area, energy consumption from the construction and operation of buildings counts for 40-50 percent of total energy consumption in Western countries.

8 __ William McDonough and Michael Braungart, *Cradle to Cradle: Remaking the Way We Make Things* (New York: North Point Press, 2002).

9 __ This development reminds us that there are two different notions of time: mechanical time and organic time. Mechanical time is the clock's measurable time, which corresponds to the conception of time that the ancient Greeks called *kronos*. Organic time is a rhythmical experience of time, which the Greeks called *kairos*. It is this last conception of time which is resuscitated in ecological construction.

10 __ Bjoern Berge, *The Ecology of Building Materials* (London: Architectural Press, 2009).

11 __ Read more about the concepts of place and space in: Claus Bech-Danielsen, *Ecological Reflections in Architecture. An Interlinked Universe Between Place and Space* (Copenhagen: Architectural Press, 2005).

12 __ Ulrich Beck, *Risk Society: Towards a New Modernity* (Barcelona: Paidos, 2006).

13 __ Robert Venturi, *Complexity and Contradiction in Architecture* (New York: Museum of Modern Art, 1966).

14 __ Christian Norberg-Schulz, *Intentions in Architecture* (Cambridge: MIT Press, 1965).

15 __ Norberg-Schulz, *Genius Loci: Towards a Phenomenology of Architecture* (New York: Rizzoli, 1979).

16 __ Jürgen Habermas, "Moderne und postmoderne Architektur" (*Arch+* 61,1982).

17 __ Wilfried Wang, "Minimalismen – en krise i samtiden." *Arkitekten* 3/98 (Copenhagen: Danish Architectural Press, 1998).

18 __ Jens Schjerup Hansen and Claus Bech-Danielsen, *Modernismens genkomst* (Copenhagen: Danish Architectural Press, 2001).

19 __ Rem Koolhaas and Bruce Mau, *Small, Medium, Large, Extra Large* (Rotterdam: 010 Publishers, 1995).

20 __ Zygmunt Bauman, *Globalisation: The Human Consequences* (New York: Columbia University Press, 1998).

21 __ Sigfried Giedion, *Architektur und Gemeinschaft* (Hamburg: Rowohlt, 1956), p. 28.

22 __ Le Corbusier, *The Home of Man* (London: Architectural Press, 1948), p. 41.

23 __ Norbert Lynton, *The Story of Modern Art* (London: Phaidon Press, 1994).

24 __ Lise Bek, "Rum er også andet end form og funktion. Renæssancens og modernismens rumopfattelse under forvandling" (*Tidens rum*. Hoersholm: Danish Building Research Institute, 1990).

25 __ Bek, *Reality in the Mirror of Art* (Aarhus: Aarhus University Press, 2003), p. 366.

26 __ Towards the end of the nineteenth century, Nietzsche had reduced God to a concept created by man, and thus the perception of nature was also changed; if God had been created by man, so had nature; that is, nature lost its divine meaning.

27 __ Clive Ponting, *A Green History of the World* (New York: Penguin Books, 1993).

28 __ Bek, *Reality in the Mirror of Art.*

29 __ For example, the modernists viewed abstraction in itself as a tool with which to liberate themselves from history.

30 __ Gevork Hartoonian, *Ontology of Construction: On Nihilism of Technology in Theories of Modern Architecture* (New York: Cambridge University Press, 1994).

31 __ Marco Frascari, "The Tell-the-Tale Detail", (*Via* 7, 1984), pp. 22–37.

32 __ This brings us back to the classic understanding of aesthetic creation as a replication of the cosmic order of nature.

33 __ James E. Lovelock, *Gaia: A new look at life on Earth* (Oxford: Oxford University Press, 1979).

Børre Skodvin

# The Complexity of Realness

To write on the subject of tectonics is a slightly intimidating task for me. Given the amount of existing scholarly writing on the understanding of tectonics in architecture, my contribution can only be based on a personal reading of how tectonic thinking might be manifested in the work of our practice.

First of all, we enjoy being involved in the process of building. This is obviously a major advantage if you are an architect, because the building process can be chaotic and disruptive and, despite all the effort that goes into the planning, it is often quite unpredictable. The architectural idea is subject to the conventionalising pressure of all the various interests with stakes in the project. As architects we need to understand and defend the limits of any possible compromise.

That being said, we have come to understand that the process of building influences and changes architecture, sometimes in profound ways. Within the academic field, some professors would argue that architecture can reach a state of ideal perfection only via the representational medium of the drawing or the model, without the necessity of transition into a full-scale physical manifestation. Our view, and indeed experience, is different. In most of our projects we find that the building process itself has an important impact on architectural content and expression. It introduces limitations and opportunities, which cannot be predicted during planning. We must respond to these disruptions in ways that do not compromise the architectural idea, but hopefully even enhance it. The speed of the process disallows long deliberation; in most cases decisions are irrevocable. Consequently, our skills as architects are put to the test. When entering competitions, we have noticed that our proposals sometimes come across as somewhat anaemic, simply because they lack the mark of the complex negotiations of the building process.

**Mortensrud Church – A Study in Tectonic Practice**

I would like to use Mortensrud Church as the basis for a discussion of our

Church interior, corner by altar

office's understanding of tectonic practice and thinking in architecture. In the making of this building, we found it possible to pursue the architectural idea to a certain level of refinement and completion, which is not at all commonplace in our experience. The client had ambitions beyond mere necessity, as well as the will and ability to embody these objectives in the process. Nonetheless, it was no simple task. The benignity of the client was counterbalanced by a contractor whose dedication to the task of construction left a lot to be desired.

After a project is finished, it is always interesting to look back and try to recall where it began. Mortensrud Church was a project quite long in the making. The client initiated the process by introducing what they called a value programme:

The building should look like a church

People should want to get married in this church

There should be a "back bench"

These three points led to a discussion that was very much focused on what the architectural character of the spaces should be like. The client expressed an explicit desire to move away from an ideal embraced in most modern Norwegian churches, where a softer egalitarianism replaces the sanctity of the traditional space for religious ceremony. At the same time, they wanted a place for ambiguity and doubt, expressed in their request for a back bench, a place for those who wish to be inconspicuously present.

There is a metaphorical language in the tradition of church building, in which interpretation and biblical and historical references permit a reading that goes well beyond the strictly architectural. Since our knowledge of this metaphorical language was virtually nonexistent, we needed to identify another platform of understanding.

### Protecting the Topography and Ecology of the Site

The appointed site for the church was a beautiful, narrow hillcrest covered with pine trees and some exposed rock. It was the kind of place where you might think it a pity to place a building. We believe that for a building to look as if it actually belongs in a specific place, it needs to connect to the site in a graceful manner. To make this connection happen, we need to know precisely what the site looks like. Beginning with a detailed topographical survey, including a mapping of every tree and its trunk diameter, we started developing an idea in which the functions of the church would be split between two buildings: one sacred space for all religious functions, and one building to house all secular activities, such as administration etc. The floor beneath the two buildings would be defined as a long rectangular base or platform positioned in a way that would allow the bedrock of the site to continue to exist underneath.

A modern building site often resembles a war zone. With all the explosives and heavy machinery, the ground is methodically destroyed and trees are cut down. We wanted to avoid this kind of mass destruction, and rather position the building with an impact that would stop as close as possible to the outside perimeter. The project was developed very much as an extension of the site itself. Geometrically speaking, the church is an addition to the existing ground. No blasting or excavation was necessary, except for the careful removal of a thin layer of soil. This technique, among other things, made it easier to preserve the existing vegetation and topography, thereby adding a dimension to the experience of the building.

Very early in the process we decided to preserve as many of the pine trees as possible. A building with large trees close to its perimeter tends to appear as if it had been there before the trees. To facilitate this conservation of the trees, the contract with the builder specified a substantial fine for each tree that was irreparably damaged during the building process. The fine was set at

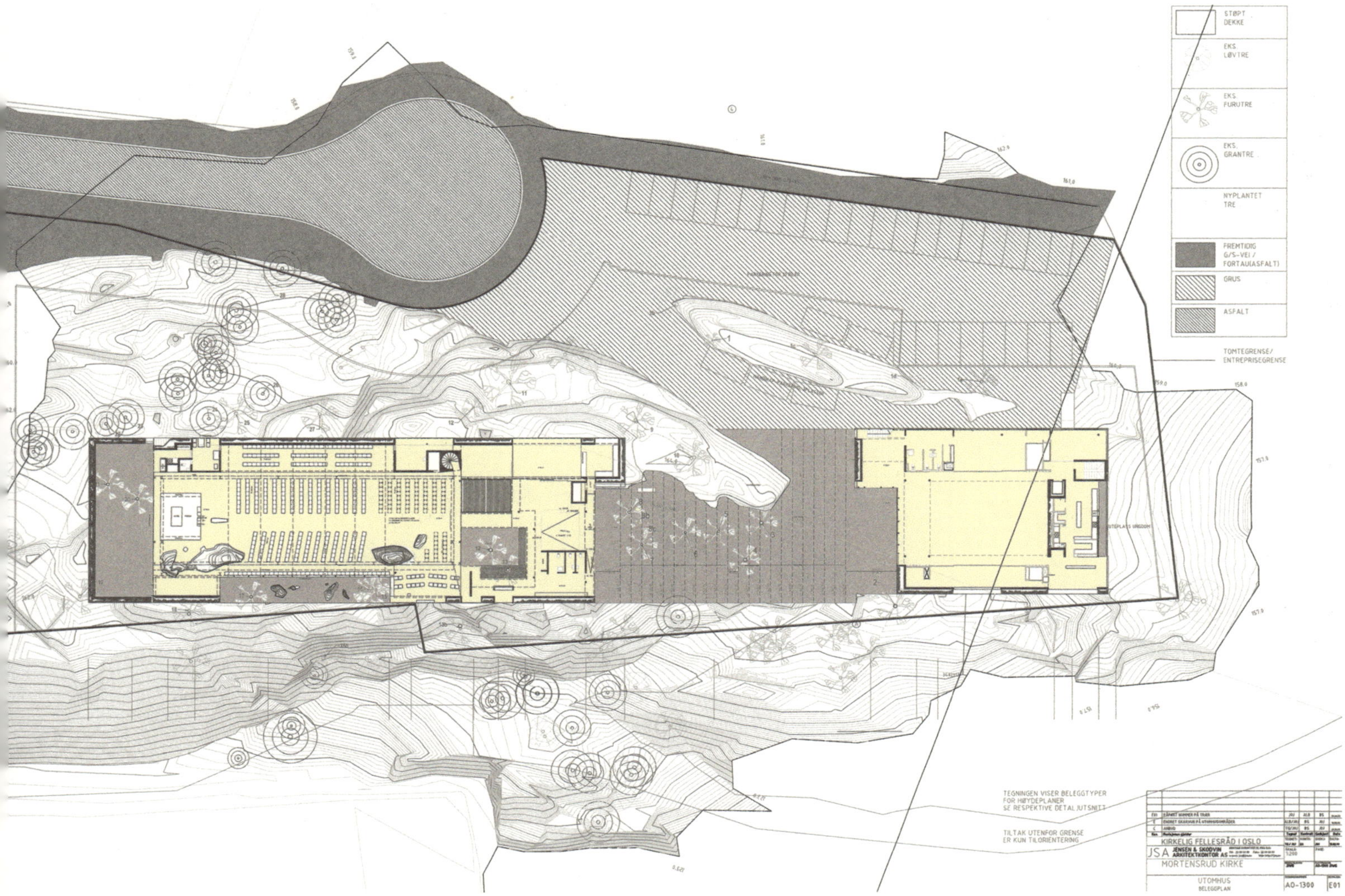

STØPT DEKKE
EKS. LØVTRE
EKS. FURUTRE
EKS. GRANTRE
NYPLANTET TRE
FREMTIDIG G/S-VEI / FORTAU(ASFALT)
GRUS
ASFALT
TOMTEGRENSE/ ENTREPRISEGRENSE

TEGNINGEN VISER BELEGGTYPER FOR HØYDEPLANER SE RESPEKTIVE DETALJUTSNITT

TILTAK UTENFOR GRENSE ER KUN TILORIENTERING

KIRKELIG FELLESRÅD I OSLO
JSA JENSEN & SKODVIN ARKITEKTKONTOR AS
MORTENSRUD KIRKE
UTOMHUS
BELEGGPLAN
A0-1300
E01

Siteplan © JSA
View of courtyard © JSA 2003

Main entrance © Jiri Havran, 2002

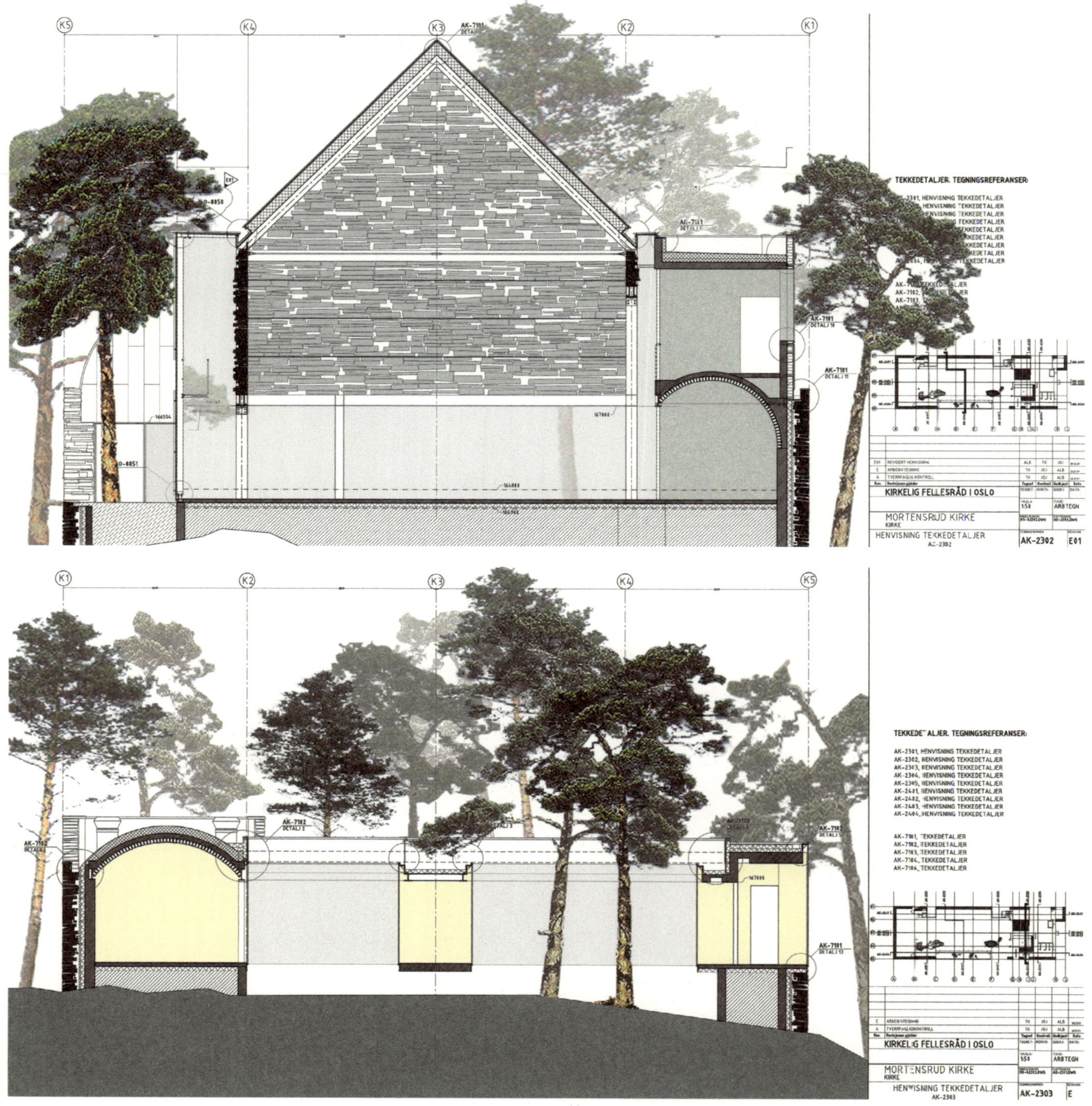


Section © JSA
Section through courtyards ©JSA

150 euros per centimetre of trunk diameter. That meant that destroying a tree with a trunk diameter of 50 centimetres would cost the builder 7,500 euros. The builder accepted these terms and, as a consequence, very few trees were lost. A number of trees are also preserved in atriums within the enclosure.

Some of the rock formations emerge like islands in the concrete floor of the church, between the congregation and the choir. Thus, the church's major divisions are a result of elements already on the site. This is possible because there are relatively large tolerances in dimensioning the rooms. No module was used to determine the exact positions of the gardens. Instead the materials and structures were chosen to facilitate a gradual and non-incremental adjustment of dimensions, without steps or modules. The tension between the wish to create a "silent", self-referential room and the various obstacles limiting this possibility was deliberately chosen as a strategy to cause architectural "disturbance" in a process in which a wide range of people and interests are involved, and which otherwise would be heavily loaded with conventional and other historical references.

### The Construction of the Inner Stone Wall

The main structure of the church is a simple steel framework with a stone wall carrying the roof. A glass façade sitting 90 – 160 cm off the stone wall defines a narrow gallery around the church room. The stone in this wall is a slate rock, and the wall is built without mortar, thus letting the light shine through. This somewhat prominent feature has become a signature mark of the building. The slate rock comes from a quarry we first visited more than 20 years ago. At that time they mainly supplied thin sheets of stone for flooring and tiles and, due to the quality requirements of these products, there was a wastage of almost 90 percent, stone which according to their standards was not fit for retail.

The flat face of this type of slate rock makes it very simple to stack. Historically it would be used to build dry mortar retaining walls or foundations. At the time of our visit, it had little or no commercial value, and was available merely for the cost of the transport. We decided to try it for some walls in a project we were working on at the time, and it turned out to be cheaper than a concrete wall. In the following years this became a very popular alternative, especially in public spaces, and the quarry could start selling all the stone, which they had previously scrapped, thus making quite a good profit.

The budget for Mortensrud Church was very tight. The church had to be built for a price per square metre equal to that of social housing in Oslo. Economically speaking, to get this building realised we had to use every possibility we could think of to get more out of less. By the time we got to build the church, the price of the slate rock had become prohibitively expensive due to popular demand. It was only possible to use it because we managed to secure a deal with the quarry,,which agreed to sell us the rock with no profit margin. We then sold it on to the contractor, thus allowing for a solution that would otherwise have been impossible.

At first glance these stone-filled walls may appear to have only a visual

Church interior © Per Berntsen, 2002

function. However, they play an important role in the structural system of the building. At 1- metre intervals, the steel columns are connected by 250-mm horizontal, flat iron straps. These act as load distributors for the stone, but at the same time, being pressed down by this considerable weight, they also act as stiffeners in the lengthwise direction of the frame structure.

Another aspect of the stone walls can be observed in the gallery between the wall and the exterior glass façade. Since the rock is quarried and not cut, it is only possible to build a wall with one even side; the other will be uneven. This uneven outside of the internal stone wall is exposed to the outside through the glass façade on three sides of the church. Although the client went quite far in arguing the need for a double-faced wall for aesthetic reasons, the budget simply would not allow it. As a consequence, the rugged outer surface remains to tell its part in the story of the building's creation. Also, it was the fact that we were working with a single-faced wall that in turn gave rise to the idea of opening up to the light between the stones, an opportunity which might otherwise have gone unnoticed.

A number of openings have been precisely cut at different, slanting angles, giving changing shapes to the sunlight as it passes across the sky. When viewed at an angle, the stone walls are closed and impenetrable to the eye. But, viewed from the front, the openly piled stones create an ambiguous connection between inside and outside. During the day, the incoming light is filtered through a filigree of stone. At night, the electric light of the inside seeps out through the spaces between the same rocks. The fact that these walls somehow contradict intuition, since the heavy mass of the stones has been placed inside a framework but this wall has not been brought all the way to the ground, has given rise to some rather unexpected criticism. Apparently, some architectural critics react negatively to what they essentially deem to be an unallowable breach of convention.

### Simple, Rough, Crafted Solutions

Another element in the planning, which was crucial in establishing architecturally satisfactory solutions, and which the budget would also allow, was to avoid conventional "proprietary" systems for façades, structures, walls, floors, etc. Rather than choosing a standard frame system for the glass façades, we designed our own, using off-the-shelf steel profiles. Not only was this solution cheaper, it also gave us far more architectural freedom. Today we see architecture around the world becoming less diverse in expression, as ever more buildings are draped in façade systems lifted from the same catalogues. The architect becomes an assembler of parts that were designed by someone else.

In the main, load-bearing structure of the steel frames, we were looking for a system of great simplicity, but with the capacity to adapt to the various obstacles imposed by the layout. At the position of the bulging rock protruding through the floor, the column splits into two separate beams, reaching the floor on either side of the rock. For the rock near the pulpit this solution was

Glass façade, gable © JSA, 2007

Church interior © JSA, 2002

Church interior, view of protruding rock © JSA, 2007

**Børre Skodvin**
founded the Norwegian office Jensen & Skodvin Architects with Jan Olav Jensen in 1995. Since then the office has gained international acclaim for their work, which includes projects like the New Monastery for Cistercian Nuns - Tautra, Mortensrud Church and Gudbrandsjuvet Landscape Hotel. Børre Skodvin is a professor at the Oslo School of Architecture, where he has also served as head of the Institute of Architecture.

not possible. Instead, the last two columns are tilted out of the lengthwise axis, to avoid landing directly on the rock. This introduces horizontal loads in the structure. These had to be compensated by introducing a compression strut in the position where the rest of the frames have only a tension stay, thereby initiating an interesting discussion. Should we allow for this anomaly, or should we correspondingly increase the dimensions of the remaining tension stays so that all members would appear similar? Both client and engineer were inclined to choose the latter option. Fortunately, we were able to convince them that this lack of similarity did not constitute a disruption of aesthetic quality, but would instead add dimensions to the experience of the users by revealing the complexity of the structural configuration.

The steel work in this building is very low-tech. It was designed to be made by a small- to medium-size workshop and did not require any specific skills or tooling. The resulting visual impression is quite rough, with the markings of the process left for all to see. Instead of paint, the steel is treated with a thin oil, which penetrates the black heat residue of the surface. We find that this makes a sympathetic impression on people, and that the visible steel inside the building is perceived as a warm material, as opposed to the more common conception that steel is a cold material.

### The Building Tells the Story of Its Own Making

By refining the design of the structural elements up to a point where they could be allowed to remain visible, we were able to avoid a procedure which has become all too common in modern architecture. Usually it is regarded as economically sound to reduce the time and resources spent on the visual quality of the structure itself. Instead, many projects choose to enclose the structural skeleton within a second envelope to conceal whatever architectural shortcomings this lack of attention might have caused. The result is actually two buildings: one to carry the loads; another to hide the structure and provide a simple, flattened surface as canvas for the architectural expression. Because we avoided this quite elaborate and costly second skin, Mortensrud Church became cheaper to build. But more importantly, the exposed structure allows for a much more interesting reading, by telling the story of its own making.

The process of designing this building took the form of a search to find a configuration that was possible, a layout that could actually be realised, given the limitations we had. It is interesting to note that the fragmented and complex character of the building that emerged from this process is virtually impossible to photograph in one shot. Instead it comes across as convoluted sequences of images, which capture only parts of the spaces and structures.

*The complete reading of the building has to be assembled in the mind of the viewer.*

# 2

# Industrialised Craft

*The transition from processes based on craftsmanship to more industrialised ways of building calls for new tectonic strategies and a re-thinking of our conception of architecture*

# Introduction

*Does contemporary industrialised construction create a new kind of coherence between parts and wholes in the organisation of the building structure, and does this support a tectonic way of thinking and practicing architecture?*

*Will the architecture of the future be an architecture of assembly?*

These questions reflect the consequences of the increasing industrialisation of today's building industry, where on-site construction is being replaced by off-site production of building elements. Prefabricated elements of various sizes and functions are then transported to the site and assembled. Unlike the industrialisation of the building processes in the period from the 1950s to the 1970s, today's industrialised building elements can easily be tailored to meet the specific needs and requirements of the architectural design, the program or the client. Mass customisation exemplifies a new business model from the product industry that uses flexible, computer-aided manufacturing systems to produce individual customised products. As a way of structuring the manufacturing processes and developing new products it has found its way into the building industry and offers prefabricated elements that are much more adaptable to various architectural designs. These manufacturing technologies introduce a new mindset, which calls for revised tectonic approaches that explore and discuss the potentials of these concepts.

It is evident that an accountable management of the natural resources needs to form the fundamental basis of the way we construct and build in the future. To meet this challenge we have to rethink the way we structure both the building processes and the final design of the construction details. Focusing on material ecologies and the understanding of the lifecycles of all elements of the building is a fundamental part of this. We need to develop constructions designed for disassembly to ensure a suitable recycling of the materials embedded in the structure. An industrial off-site production of the different parts of the building enables a high degree of control of the flow of materials. This can reduce the waste of material within the process, but new tectonic strategies have to define methods and details linking an industrial way of production with an ecological approach to the management of resources.

Another important aspect of an ecological way of thinking is the reduction of the consumption of energy from nonrenewable resources. This involves new technologies that need to become an integrated part of the building design. It is difficult to foresee future technologies. Consequently building structures should be able to adapt to and accommodate unpredictable situations and elements.

Future tectonic strategies need to include adaptability and disassembly in the development of industrialised solutions, which might create a new kind of coherence between parts and wholes. As a result of this the architecture of the future could be montage, assembly or assemblage architecture – creating mutable wholes out of autonomous parts.

**Positions and Perspectives**

To discuss the challenges of the transition from processes based on craftsmanship to more industrialised ways of building, five different approaches are presented in the following chapter. The first text, by Fredrik Nilsson, is a general and theoretical introduction to the central challenges in contemporary construction and architecture. This discussion is followed by more specific case studies introduced by Ulrik Stylsvig Madsen, discussing the need for new aesthetic values in the field of architecture. The practice of Lundgaard & Tranberg is presented by Peter Thorsen in the third text as an example of how to work with tectonic principles in contemporary construction. The fourth text, by Ole Egholm Pedersen, is based on new tectonic strategies for the casting of concrete. Finally, the concept of montage introduced by Charlotte Bundgaard discusses new principles for architecture based on prefabricated elements.

The five texts can be recapitulated in the following short descriptions of the different perspectives:

In the text "Making as Architectural Thinking and Practice" the Swedish architect and researcher Fredrik Nilsson introduces the aspect of "making" – actively doing, modelling, and altering things – as a central aspect of architecture, both with regard to the actual process of designing and to how knowledge is used and generated. New digital technologies and means of production influence the architectural profession and ways of working in both conceiving and producing built environments. This text discusses how these challenges in the contemporary situation call for rethinking central concepts in architecture, and how we need to reformulate the ontological and representational aspects as well as material and expressive relations. In this discussion theories and concepts by Manuel DeLanda, Kenneth Frampton and Stan Allan, etc., are used as argumentation for articulating and reconnecting the core characteristics of architectural practice and making.

"Constructing Immediacy" is the title of the text by the Danish architect and researcher Ulrik Stylsvig Madsen. He argues that an ecological way of thinking links the well-being of the individual to the preservation of the individual's surrounding environment as a whole. Understanding and conforming to this adds a new ethical dimension in the way we think and construct architecture. In this text the concept of "immediacy" is introduced as a tectonic strategy to connect ecological principles with the construction of buildings. The aim is to discuss the increased complexity particularly in the context of industrialised building processes by suggesting a new relation between parts and wholes that challenges the traditional aesthetic values in the field of architecture. The discussion is based on a comparative case study of Can Lis (Jørn Utzon) and Loblolly House (KieranTimberlake).

In the text "Practice as a Constructional Craft!" the Danish architect Peter Thorsen introduces the term "constructional craft" as a way of describing the work of the Danish architectural office Lundgaard & Tranberg Architects – a work focusing on the knowledge of building technologies and processes in order to achieve tectonic clarity in both individual projects and as a contribution to innovation and further development of the field of architecture. Based on a regional and contextual approach to the design process, the office investigates the inherent relationship between form, content, construction and materials – and the possible integration of these into a seamless whole, turning every project into a resolved architectural statement. In the text an exhaustive exposition of the Tietgen Dormitory project forms the basis of a description of the work methods and key values of the office.

Based on the PhD thesis of Danish architect and researcher Ole Egholm Pedersen, the text "The Tectonic Complex of Concrete" introduces a tectonic approach to complex forms in concrete and proposes tectonics as a strategy for supporting the development of individualisation and resource optimisation in concrete casting. Problems with existing techniques for casting complex shapes in concrete give rise to a case study in which a novel concrete casting technique is developed based on the technique of folding. The case study gives rise to the conclusion that the logic of the technique should be a determining factor for the generation of form rather than a means to realise a preconceived form. This will reduce fabrication time and enhance tectonic qualities insofar as the logic of technique is clearly expressed in the geometric form.

In "Ready-mades Revisited" the Danish researcher and architect Charlotte Bundgaard explores current challenges of industrialisation in architecture. The text intends to discuss the impact and potentials of tectonics and sustainability when dealing with prefabricated building components and juxtaposition. The concept of "montage" is proposed as a means for investigating possible strategies that might point to fruitful crossovers between tectonics, sustainability and industrialisation. The discussion is based upon theoretical positions and a study of the work of the French architectural office Lacaton & Vassal.

Ulrik Stylsvig Madsen

Fredrik Nilsson

# Making as Architectural Thinking and Practice

The aspect of "making", of actively doing, modelling, altering things, is a central aspect of architecture, both in terms of the actual process of designing and of how knowledge and theories are used and generated. Making is part of how architects and designers think and how knowledge is generated and used. This point of view is a way of connecting and managing external factors in the constantly changing conditions architecture is involved in. Artefacts play a central role in this concept, both as bearers of knowledge and as results of processes of making. The act of making is a way to work with and integrate all the different perspectives which characterise architecture, and which have to be dealt with. It is an integral part of a design ability used to generate entities of contradictory, heterogeneous elements and aspects in the production of objects and artefacts.

The built environment consists of materialised knowledge of many kinds, which can be found, for instance, in the detailing, structure and assemblage of spaces and materials in vernacular architecture, as well as in contemporary design objects and infrastructural projects. It is possible to read these different types of knowledge in a building on various levels: e.g., as patterns of social and cultural processes as well as material and technical patterns and mechanisms. New digital tools have provided more possibilities for visualising and reading these patterns, but also for integrating knowledge in other ways, not least architectural and engineering perspectives on structural technologies with new possibilities for expression and precision in what is built. One central concept in architecture related to this is, of course, the concept of tectonics.

During recent decades Kenneth Frampton has been a central figure in discussions on tectonics. He describes tectonics as *a poetics of construction*, where the full tectonic potential of a building comes from its capacity to articulate both the poetic and the cognitive aspects of its substance. With reference to Semper's distinction between symbolic and technical aspects of building, Frampton makes a distinction between the *representational* and *ontological* aspects of tectonic form. He also states that this dichotomy must

The exhibition *Beauty and Waste* by Herzog & de Meuron showed the physical traces of design and thought processes that projects leave behind, in the form of models, sketches and material samples. 2005. © Fredrik Nilsson

constantly be rearticulated in the creation of architectural form, since every building type, technology, topography and temporal circumstance generates different cultural situations and conditions.[1] The call, also by Frampton, for a constant reformulation of the aspects of tectonics is something to take into serious consideration in contemporary developments.

Anne Beim and her colleagues of the research project "Towards a Tectonic Sustainable Building Practice" have stressed that tectonic thinking is principally concerned with the making and the application of building materials when constructing architecture, and that this concern becomes a creative force when designing and construing architecture. In their view, tectonic thinking can be used to identify and refine strategies for developing and improving the contemporary building industry.[2] In my view, the elaboration, constant reformulation and articulation of central concepts of architecture, of which the tectonic is one, is also crucial for the development of the discipline, and the specific knowledge of architecture and its practice.

Architecture has been described as a "making discipline"[3] and as a "material practice"[4] in attempts to develop more field-specific scholarship and to articulate the characteristics of that field. Central here is a certain "making knowledge", which relates to knowing how to do things based within the profession, and aspects of transforming reality based on contemporary conditions, but with an eye to the future. The discipline's foremost concerns are performances and transformations rather than representations and interpretations of a past. But what are the relations between the notions of making and of practice? How can they be considered in contemporary building production? What kinds of mechanisms and processes are involved when conceiving and producing architecture, and with what frameworks could tectonics be rethought today? This text is a brief attempt to reflect on some of these questions.

Architectural designs as explorative material practices related to the processes of making have been a recurring theme at the Architecture Biennale in Venice. Image right shows *Ungapatchket* by Gehry Partners in Venice 2008, and image below shows "Work-Place" by Studio Mumbai in Venice 2010. © Fredrik Nilsson

### The Production and Practice of Architecture

Aristotle had one concept for the theoretical, but two for the practical. He distinguished between a *productive activity* (*poíêsis*), which leads to something produced (material or immaterial), and a *practical activity* (*praxis*) pursued for its own sake and having a value in itself.[5] For Aristotle certain human activities or forms of life lead to certain abilities or forms of knowledge. You acquire scientific skills and knowledge (*epistêmê*) by engaging in theoretical activity (*theôría*). Practical and technical knowledge (*téchnê*) is acquired through the activities of making and production (*poíêsis*), and practical wisdom (*frônésis*) is acquired through actions and by acting (*praxis*). The distinction between poíêsis and praxis implies a distinction between instrumental actions, which are performed to reach a goal (e.g., to build a house) and actions which are meaningful, such as self-fulfilment or the satisfaction of a life need (e.g., to dwell). Making and the productive activity (poíêsis) imply a process that takes a certain time before the goal is reached; the goal and the process (the instrumental activity) do not exist simultaneously. The meaningful action (praxis) is not a process, but the actual goal; the meaning and the action are the same.[6]

In Aristotle's thinking there is a certain relation between making and practice, between instrumental and meaningful actions (between poíêsis and praxis), which is not clearly seen today. Poíêsis implies the productive actions of making, doing something, which could also lead to immaterial artefacts, such as an organisation or a theory. Praxis implies actions with broader implications and ethical dimensions that are important in a profession, and the role they play in a community or society. In a concept like tectonics one finds interesting traces of both, since it is about simultaneously fulfilling instrumental tasks and expressing meaning or evoking experiences.

The architectural language of Atelier Bow-Wow is a result of dialogues between clients' lifestyles, site conditions and construction technology, as investigated in the models of "House Behaviorology".
© Fredrik Nilsson

NEW YORK
LOS ANGELES
SHANGHAI
SHANGHAI

Architecture as connections and assemblages of heterogeneous aspects and elements. Project presentation © White Arkitekter

In an essay on the characteristics of the discipline of architecture, David Leatherbarrow has noted that Aristotle distinguished three sorts of knowledge: technical, ethical and philosophical. They correspond to three types of activities characterising human life: production, action and contemplation. "The result of production is something made, of action something done, and of contemplation something envisaged or desired."[7] What architects normally make is, as many have argued, not buildings, but rather representations of buildings, cities and rooms or some of their details. But here the kind of "representation" is less mimetic than prospective, because architectural design has always been separated from production. Unlike many craftsmen and tailors, architects do not make the things they design. They make "design information", an equivalent of making a pattern of clothing, with which the architect simultaneously forms the immaterial host – the space – and the material draping or container for it. Since the introduction of digital fabrication technologies, the conventional protocols of exchange between design and making have been thoroughly redefined, putting them in other, more integrated relations, which change the negotiated translations between steps in the process.[8] Leatherbarrow also notes that architectural representations normally come in sets, in which each representation is "cross- referenced" to many

Physical models showing the density and spatial character of cities as assembled wholes at the 10th Architectural Biennale *Cities. Architecture and Society*, 2006. © Fredrik Nilsson

DENSITÀ
LOS ANGELES
CAIRO
SHANGHAI

others, forming a complex set of relations between representations of different kinds. "Architectural understanding means grasping a network, weave, or matrix of figures, each partial but all mutually dependent."[9] Architectural representations, Leatherbarrow argues, also show aspects of the world, which are "hidden" to non-architects, thus disclosing aspects that otherwise would be unseen. The chief skill of the architect is to master the craft of making certain representations, and here one could add that what the architect does is to make these "hidden" relations visible, to construct the relations between aspects and elements through certain artefacts.

When Albena Yaneva has followed architects in their "architectural laboratories" it is mainly to study such relations or "associations", where she investigates concrete architectural objects, networks, moves and habits in architectural practices. Here design invention and form come from sets of everyday trajectories of models and persons in the office space, experiments with materials and mock-ups, and presentations for clients and users, all of which transform and leave traces in the architecture. Yaneva identifies and points to the immense capacity of design objects to connect heterogeneous actors.

*"This particular capacity of a building to associate both human and non-human actors, and in different periods of time, makes it an important social actor."*[10]

She also states that a building is less an artefact, a construction, or conceivable as a modernist object. It is rather to be conceived as a "thing" in the terms of Bruno Latour, "a contested assemblage".

Architectural thinking and making are about associating, connecting elements (both non-human and human, material and immaterial) via the activities of modelling, testing and transforming artefacts that are often material and involve material thinking. But how are we to understand architectural making in the contemporary technological situation, in this world of flows, networks, digital information and transient connections?

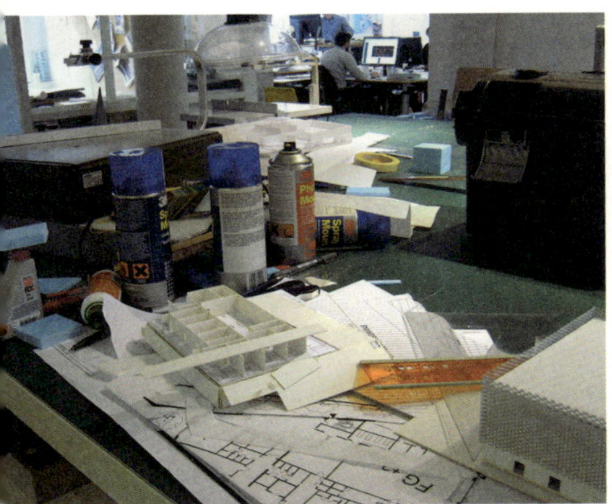

Architectural offices as laboratories investigating, relating and assembling constructions, spaces and people through different artefacts, materials and representations. White arkitekter, Göteborg, Sweden, 2010. © Fredrik Nilsson

#### Wholes as Assemblages of Parts

An important point of reference in recent discussions is the notion of assemblage theory as developed by Manuel DeLanda, mainly building upon and very much influenced by the thinking of Gilles Deleuze. Assemblage theory is meant to apply to a wide variety of wholes constructed from heterogeneous parts. Entities, ranging from atoms and molecules to biological organisms, species and ecosystems may, according to DeLanda, be usefully treated as assemblages.[11] This theory can also be applied to social entities, thus making it possible to cut across the nature-culture divide or open it up into a continuum to be analysed in other ways.

To make possible the analyses of wholes, which are at the same time irreducible and decomposable, DeLanda uses the concept of *assemblages*.[12] Seminal to assemblage theory is the fact that the synthesis of the properties of a whole is not reducible to its parts. DeLanda contrasts Hegelian totalities, in which the parts form a seamless whole, an organic unity, with assemblages, in which the parts do not form a seamless whole. Assemblages are rather wholes whose properties emerge from interactions between distinct parts.[13] Here there are different kinds of relations between parts, and DeLanda talks about *relations of interiority* and *relations of exteriority*.

A theory of totalities is based upon the concept of *relations of interiority*, where the component parts are constituted by their relation to other parts in the whole. A part detached from such a whole ceases to be what it is, since being that particular part is a constitutive property of the part. In contrast, assemblages are wholes characterised by *relations of exteriority*. These relations imply that a component part is not defined by its relations, and a

10th Architectural Biennale, *Cities, Architecture and Society*, Venice, 2006. © Fredrik Nilsson

part of an assemblage may be detached from it and plugged into a different assemblage, where its interactions are different. A theory of assemblages and the exteriority of relations imply a certain autonomy for the parts, and the properties of a whole are not the result of an aggregation of the components' properties, but of the actual exercise of their capacities.

*"Relations of exteriority guarantee that assemblages may be taken apart while at the same time allowing that the interactions between parts may result in a true synthesis."*[14]

The heterogeneity of components is also an important characteristic of assemblages. The complex interactions between component parts are crucial for the *emergence* of properties of the whole.

In addition to the exteriority of relations, DeLanda argues that the concept of assemblage is defined along two dimensions or axes.[15] One of the axes is the variable material and expressive roles that the component may play in the assemblage: from a purely material role at one end of the axis to a purely expressive role at the other. These roles are variable and may occur in mixtures, and a component may play a combination of material and expressive roles by exercising different sets of capacities. The other axis is the variable stabilising and destabilising processes in which the component becomes involved. At one end are the processes that stabilise the identity of the assemblage by increasing internal homogeneity or sharpening its boundaries, and at the other are processes that destabilise the identity by increased heterogeneity or blurred borders.

While in his book *A New Philosophy of Society* DeLanda analyses mainly social entities and assemblages, he also analyses buildings and the built environment on different scales. He looks at individual buildings, neighbourhoods and cities as assemblages. He discusses the material role played by components: e.g., load-bearing structures and connectivity of spaces, and how assemblages have changed in history as a result of new building technologies such as reinforced concrete, steel structure and inventions like escalators, elevators and ventilation systems. He also discusses the expressive roles of components, such as façades, spatial form, furniture, decorative treatment of walls, floors, ceilings, and how these physical expressions often go together with linguistic expressions (e.g., like Gothic churches), forming various assemblages.[16] Assemblage theory, with its axes of material and expressive roles, and processes of stabilisation or destabilisation of identities, could make a fruitful contribution when developing the theoretical framework of tectonics. For example, there are interesting parallels to be drawn with the notions of how wholes are formed or emerge from the interactions of parts.

### The Digital Crafting of Architecture

Regarding how to construct something that works and is experienced as an architectural whole, Leatherbarrow has argued that technological knowledge lacks a sense of wholeness or concern for it. Standardised building systems

Tests and presentations of prototypes for Green Façade building system. R&D project by White arkitekter and NCC. Prototypes at White's office in Stockholm, 2012. © Fredrik Nilsson and White

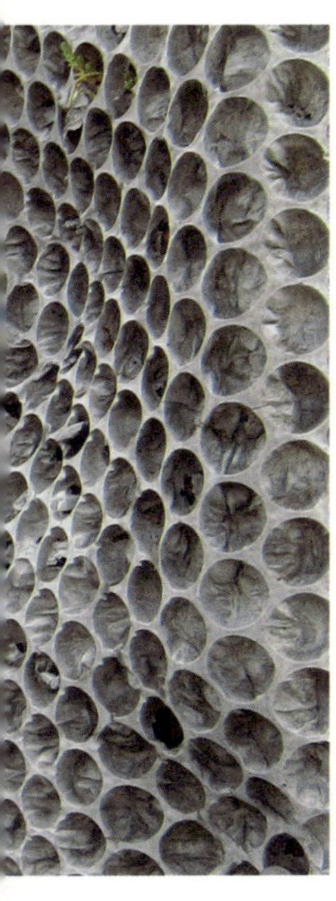

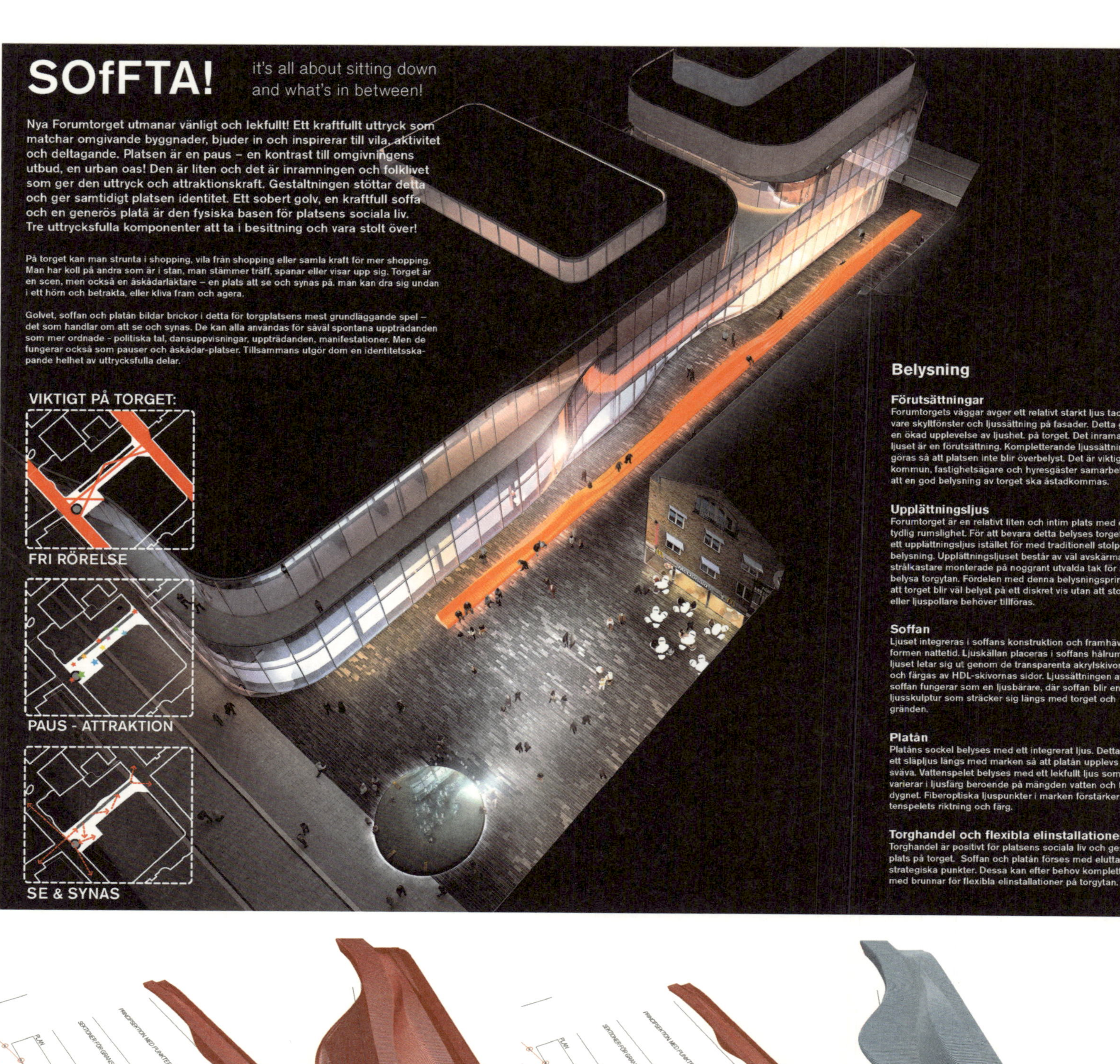
SOfFTA!
it's all about sitting down
and what's in between!
Nya Forumtorget utmanar vänligt och lekfullt! Ett kraftfullt uttryck som matchar omgivande byggnader, bjuder in och inspirerar till vila, aktivitet och deltagande. Platsen är en paus – en kontrast till omgivningens utbud, en urban oas! Den är liten och det är inramningen och folklivet som ger den uttryck och attraktionskraft. Gestaltningen stöttar detta och ger samtidigt platsen identitet. Ett sobert golv, en kraftfull soffa och en generös platå är den fysiska basen för platsens sociala liv. Tre uttrycksfulla komponenter att ta i besittning och vara stolt över!
På torget kan man strunta i shopping, vila från shopping eller samla kraft för mer shopping. Man har koll på andra som är i stan, man stämmer träff, spanar eller visar upp sig. Torget är en scen, men också en åskådarläktare – en plats att se och synas på. man kan dra sig undan i ett hörn och betrakta, eller kliva fram och agera.
Golvet, soffan och platån bildar brickor i detta för torgplatsens mest grundläggande spel – det som handlar om att se och synas. De kan alla användas för såväl spontana uppträdanden som mer ordnade - politiska tal, dansuppvisningar, uppträdanden, manifestationer. Men de fungerar också som pauser och åskådar-platser. Tillsammans utgör dom en identitetsskapande helhet av uttrycksfulla delar.
VIKTIGT PÅ TORGET:
FRI RÖRELSE
PAUS - ATTRAKTION
SE & SYNAS
Belysning
Förutsättningar
Forumtorgets väggar avger ett relativt starkt ljus tack
vare skyltfönster och ljussättning på fasader. Detta ge
en ökad upplevelse av ljushet. på torget. Det inramanc
ljuset är en förutsättning. Kompletterande ljussättning
göras så att platsen inte blir överbelyst. Det är viktigt
kommun, fastighetsägare och hyresgäster samarbetar
att en god belysning av torget ska åstadkommas.
Upplättningsljus
Forumtorget är en relativt liten och intim plats med en
tydlig rumslighet. För att bevara detta belyses torget m
ett upplättningsljus istället för med traditionell stolp-
belysning. Upplättningsljuset består av väl avskärmade
strålkastare monterade på noggrant utvalda tak för att
belysa torgytan. Fördelen med denna belysningsprinci
att torget blir väl belyst på ett diskret vis utan att stolp
eller ljuspollare behöver tillföras.
Soffan
Ljuset integreras i soffans konstruktion och framhäver
formen nattetid. Ljuskällan placeras i soffans hålrum c
ljuset letar sig ut genom de transparenta akrylskivorna
och färgas av HDL-skivornas sidor. Ljussättningen av
soffan fungerar som en ljusbärare, där soffan blir en
ljusskulptur som sträcker sig längs med torget och
gränden.
Platån
Platåns sockel belyses med ett integrerat ljus. Detta g
ett släpljus längs med marken så att platån upplevs at
sväva. Vattenspelet belyses med ett lekfullt ljus som
varierar i ljusfärg beroende på mängden vatten och ti
dygnet. Fiberoptiska ljuspunkter i marken förstärker va
tenspelets riktning och färg.
Torghandel och flexibla elinstallationer
Torghandel är positivt för platsens sociala liv och ges
plats på torget. Soffan och platån förses med eluttag
strategiska punkter. Dessa kan efter behov komplette
med brunnar för flexibla elinstallationer på torgytan.

need to adapt to the situation, and the adjustments of standard elements result from the architect's understanding of how all aspects come together to give durable dimension and shape to the patterns of our lives. Leatherbarrow states that architects can avoid neither craft nor industry and must develop an intuitive grasp of manual procedures and scientific understandings of the world, so they can manage predictions about performances of elements.

*"In the terrain called technology, no fork in the road demands a choice between craft and industrial methods; instead of assuming or mapping out a divergence, we must discover and describe a convergence; we need to see how manual and conceptual technologies intersect with one another along the lines of a unified understanding of building production."*[17]

Architecture as a material practice implies that making, and the close engagement with materials are intrinsic to the design process. Branko Kolarevic has stressed that this making is increasingly mediated through digital technologies. The *digital making*, through the use of digital tools in design and material production, is blurring the past century's sharp boundaries between conception and production. Parametric design and digital fabrication are restructuring the relations between design and production, enabling the closer interrogation of materials during early design phases. In this context, Kolarevic argues, the craft no longer lies only in production, but also in the definition and manipulation of geometry and in the engagement of materials and production in the feedback loops of digital design processes.[18]

Kolarevic also refers to David Pye and his definition of craftsmanship from 1968 as particularly suitable for the contemporary "digital age". Pye writes:

*"If I must ascribe a meaning to the word craftsmanship, I shall say as a first approximation that it means simply workmanship using any kind of technique or apparatus, in which the quality of the result is not predetermined, but depends on the judgment, dexterity and care which the maker exercises as he works. The essential idea is that the quality of the result is continually at risk during the process of making; and so I shall call this kind of workmanship 'The workmanship of risk': an uncouth phrase, but at least descriptive."*[19]

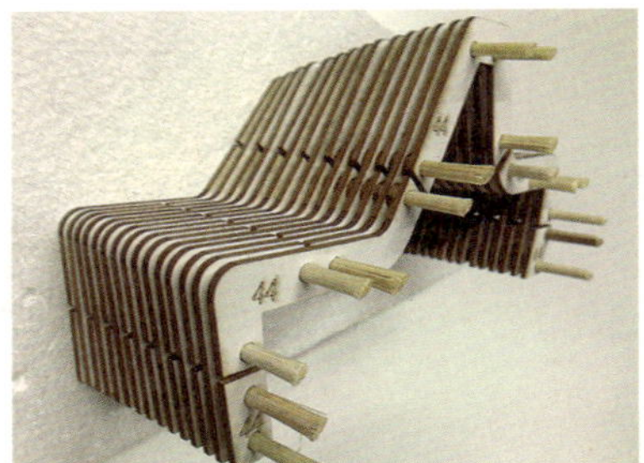

Development of design of Forumtorget, Uppsala, Sweden, through parametric and material modelling. © White arkitekter, 2012.

Pye also contrasts the workmanship of risk with the workmanship of certainty. The latter is to be found in quantity production and, in its purest state, in full automation. Here the quality of the result of each operation conducted during production is exactly predetermined.

The "craftsmanship of risk", in which the outcome "is continually at risk", has resonance today, Kolarevic argues, when many practitioners have adopted digital technologies to discover new, promising configurations, spatial potentials and organisations. These are not predetermined, but "crafted" through generative procedures, where digital media are not only representational tools for visualisation, but generative tools. Parametric variation can be automatic or controlled manually by discrete, incremental steps. The descriptions of relations between objects become the structuring, organisational principles for generation and transformation.

*"How these interdependencies are structured and reconfigured depends to a considerable extent on abilities of the designer to craft these relationships precisely. In parametric design, the conceptual emphasis shifts from particular forms of expression (geometry) to specific relations (topography) that exist within the project."* [20]

The capacity of parametric techniques to generate new opportunities is highly dependent on the perceptual and cognitive abilities of the designer, since the discovery and recognition of emergent forms and potentials is based in qualitative cognition.

*"Just like craftsmen of the past, the craftsmen of the digital age – the designer working with virtual representations of the material artefacts – seek out*

*unpredictable outcomes by experimenting with what the medium and the tools have to offer,"* and Kolarevic continues: *"Knowing what, why, and how to adjust requires deep knowledge of the processes, tools, and techniques, just as it did in the pre-digital era."*[21]

But to inform potential architecture by tectonic means of production, the processes often need physical prototyping as part of the feedback loop and critical investigation of the quality of construction.

Michael Stacey has argued that the haptic is important in the link between thinking and making, but once drawing, sketching and making by hand have been learnt, many issues can be explored completely digitally. But he stresses that although

*"a totally cerebral digital crafting of architecture is possible, in this process the prototype becomes an even more important tool. A flip-flop from the digital to the physical and back again stimulates thinking about design, the rigorous thinking through of architecture".*[22]

In Stacey's view, the externalisation of ideas is crucial, and the integration of making into the design process is more important than whether or not haptic means are employed.

**Reconnecting to the Core Characteristics of Architectural Making**

In conclusion one can say that architectural practice and making are also very close, and perhaps even more, in the contemporary digital world. New digital technologies and means of production influence the architectural profession and ways of working in both conceiving and producing built environments. These changes and the contemporary situation call for the rethinking of central concepts in architecture, not least the tectonic and its related conceptual frameworks, where we need to reformulate the ontological and representational aspects as well as the material and expressive relations. DeLanda's notions of how wholes are formed, and the dimensions of variable material and the expressive roles that components play in assemblages, may support the elaboration of theories of tectonics.

But one can also see that DeLanda's assemblage theory connects to earlier reflections in architecture. Before the broad invasion of digital tools, Christian Girard wrote about "nomadic concepts" as central and characteristic of architectural thinking. Nomadic concepts "keep infringing on extra-linguistic grounds, getting entangled with affects, drives, materials which do not obey the so-called 'rules' of language and do not form a 'system'". Girard continues:

*"Contrary to what can often be read, the design process cannot be compared to synthetizing elements of knowledge called upon by the architect: such an idealistic view is akin to a myth. Nomadic concepts do not produce a synthesis: they only make it possible, in a transient way, punctually, for heterogeneous levels of reality to combine with each other, to conglomerate, into a new dimension or 'plateaux', to use the terminology of Deleuze and Guattari."* [23]

So there are reasons to consciously reconsider and relate to parts of architectural thinking and concepts from different periods of history, but constantly to rearticulate them in relation to current cultural and technological situations. The conscious and critical consideration of theoretical frameworks, such as assemblage theory and actor-network theory, relating to contemporary material and societal conditions, also give opportunities to articulate and reconnect to the core characteristics of architectural practice and making.

**Fredrik Nilsson**
is professor of architectural theory at Chalmers University of Technology and also head of research & development at White Arkitekter in Sweden. In this way he combines research with architectural practice. He is currently director of the research environment/project "Architecture in the Making: Architecture as a Making Discipline and Material Practice" funded by the Swedish Research Council Formas 2011–2016. Fredrik Nilsson holds a PhD in Architecture from Chalmers University of Technology, and recently contributed to the book *Design Innovation for the Built Environment – Research by Design and the Renovation of Practice* (2012). He is currently editing the book *The Changing Shape of Practice. Integrating Research and Design in Architectural Practice* together with Michael Hensel (forthcoming in 2014).

Façade for Stockholm Tele2 Arena in different kinds of materiality and models. © White arkitekter, 2000-2013.

1 __ Kenneth Frampton, *Studies in Tectonic Culture: The Poetics of Construction in Nineteenth and Twentieth Century Architecture* (Cambridge, Mass: The MIT Press, 1995), p. 16.

2 __ Claus Bech-Danielsen, Anne Beim, Karl Christiansen, Charlotte Bundgaard, et al., *Tectonic Thinking in Architecture* (Copenhagen: The Royal Danish Academy of Fine Arts, 2012), p. 12.

3 __ Halina Dunin-Woyseth and Jan Michl, eds., *Towards a Disciplinary Identity of the Making Professions* (Oslo: AHO, 2001).

4 __ Stan Allen, *Practice: architecture, technique and representation* (Amsterdam: G+B Arts International, 2000).

5 __ José Luis Ramírez, *Skapande mening. En begreppsgenealogisk undersökning om rationalitet, vetenskap och planering* (Stockholm: Nordplan, 1995), p. 8.
See also Aristotle, *Nichomachean Ethics* (Kitchener: Batoche Books, 1999), bk. 6.

6 __ José Luis Ramirez, *Designteori och teoridesign* (Stockholm: Nordplan, 1995), p. 7.

7 __ David Leatherbarrow, "Architecture Is Its Own Discipline," in *The Discipline of Architecture*, ed. Andrzej Piotrowski and Julia Williams Robinson (Minneapolis: University of Minnesota Press, 2001), p. 85.

8 __ Bob Sheil, "Introduction," in *Manufacturing the Bespoke. Making and Prototyping Architecture* (Chichester: John Wiley & Sons, 2012), pp. 6-9

9 __ Leatherbarrow, "Architecture Is Its Own Discipline," p. 88.

10 Albena Yaneva, *The Making of a Building. A Pragmatist Approach to Architecture* (Bern: Peter Lang, 2009), p. 198.

11 __ Manuel DeLanda, *A New Philosophy of Society. Assemblage Theory and Social Complexity* (London: Continuum, 2006), p. 3.

12 __ Manuel DeLanda, *Philosophy and Simulation. The Emergence of Synthetic Reason* (London: Continuum, 2011), pp. 184-185.

13 __ De Landa, *A New Philosophy of Society*, pp. 4-5.

14 __ De Landa, *A New Philosophy of Society*, p. 11.

15 __ De Landa, *A New Philosophy of Society*, pp. 9-11.

16 __ De Landa, *A New Philosophy of Society*, pp. 96-100.

17 __ Leatherbarrow, "Architecture Is Its Own Discipline," p. 101.

18 __ Branko Kolarevic, "The (Risky) Craft of Digital Making," in *Manufacturing Material Effects. Rethinking Design and Making in Architecture*, ed. Branko Kolarevic and Kevin Klinger (New York: Routledge, 2008), p. 120.

19 __ David Pye, *The Nature and Art of Workmanship* (Cambridge: Cambridge University Press, 1968), p. 20.

20 __ Kolarevic, "The (Risky) Craft of Digital Making," p. 121.

21 __ Kolarevic, "The (Risky) Craft of Digital Making," p. 127.

22 __ Michael Stacey, "Digital Craft in the Making of Architecture," in *Manufacturing the Bespoke. Making and Prototyping Architecture*, ed. Bob Sheil (Chichester: John Wiley & Sons, 2012), p. 76.

23 __ Christian Girard, "The Oceanship Theory. Architectural Epistemology in Rough Waters," *Philosophy & Architecture. Journal of Philosophy and the Visual Arts*, no. 2 (1990), p. 80.

Ulrik Stylsvig Madsen

# Constructing Immediacy

An ecological way of thinking links the well-being of individuals to the preservation of their surrounding environment as a whole. Understanding as well as conforming to the environment adds a new ethical dimension in the way we think and construct architecture. This need for rethinking the way we construct our physical surroundings should be seen in the light of the industrialisation of building processes. This leads to the principal question of this text:

How can we accommodate the increased complexity in the way we build by introducing a new relation between parts and wholes within the building structure, and will this new relation challenge traditional aesthetic and cultural values within the field of architecture?

**Tectonics and Ecology – the Correlation between Parts and Wholes**

This text focuses on one of the key concepts within the field of tectonics: the correlation between parts and wholes in the building structure. Many theoreticians, from Gottfried Semper to those of modern times, have described this relationship. Currently, one of the most important texts focusing on this aspect of tectonics is *The Tell-the-Tale Detail* by the Italian architect and writer Marco Frascari. In the text, the role of the detail is defined as follows:

*"That is to say the 'construction' and the 'construing' of architecture are both in the detail. [...] Details are much more than subordinate elements; they can be regarded as the minimal units of signification in the architectural production of meanings."*[1]

This way of thinking merges the way architecture is constructed as a corporeal and process-related structure with the way the perceiver can construe a certain meaning from the final result. The individual detail becomes the smallest element in the construction of the significance of the building as a whole. Within a Nordic context, the Danish architect and researcher Anne Beim has described the same problem this way:

*"This essential task of architecture also points towards the very heart of tectonic endeavour, defined as a holistic attitude consisting of signifying*

Can Lis – view of the passage between the living room unit and the bedroom unit. © Ulrik Stylsvig Madsen

*interrelated distinguishable parts or phenomena that form the sum total of the building as an aggregate."*[2]

This emphasises the perception of the structural logic of the building as the very core element of tectonic practice. In other words, the tectonic dimension of a work of architecture is the possibility for the perceiver to read and understand a conscious and intentional processing of the construction. Every detail has its own specific design which refers to the way the different elements within the structure are interrelated. In this way, the logic of the construction as a whole expresses how the different parts interact to form the overall structure.

As outlined both in the introduction to this anthology and in the text by David Leatherbarrow, the scientific field of ecology was founded within the natural sciences of the 19th century. The concept derives from the Greek term *oikos*, which signified the household, combined with the term *logos*. Ecology would then mean something like a principle of order of the household. The shortest definition of the field of ecology would be a study of interactions among organisms and their environment. An ecological understanding of the world thus focuses on the mutual dependency between the life of the individual (parts) and the natural systems (wholes) in which this life is embedded. This links back to the etymological origin of the word, pointing at the management of the household, a situation in which the individual bears responsibility for the well-being of the overall structure. If we as humans are to take this responsibility seriously in the way we manage the ecosystems of our surrounding environment, we may need to rethink the guiding principles of the way we structure our lives. How this ethical challenge affects the way we think and the way we design architecture with a special focus on the tectonic dimension will be the focus of the following discussion.

### An Ecological Approach to the Perception of Tectonics

The essence of a tectonic approach to architecture is that the logic of the construction of the building can be perceived and interpreted by the user of the building. This is highlighted in the following extract from Anne Beim's *Tectonic Visions in Architecture*.

*"Signification in construction or the tectonic dimension in architecture depend on how 'the creator' (the architect), as well as 'the spectator' (the user), perceives and interprets constructions as well as what sort of meaning (or lack thereof) they transfer into the physical solutions of an architectural project."*[3]

This transfer of meaning from the building to the perceiver might seem a simple process, but it is rather complex. It could be argued that the meaning does not arise before the perceiver construes it, and that this construing depends solely on the knowledge and experience of this individual. Another view of this process of perceiving was introduced by the American psychologist James J. Gibson, in his development of the concept of "affordances". Rooted within the field of ecology, this concept introduces a direct relationship between the object and the perceiver. Gibson describes it in this way:

*"Perhaps the composition and layout of surfaces constitute what they afford. If so, to perceive them is to perceive what they afford. This is a radical hypothesis, for it implies that the 'values' and 'meanings' of things in the environment can be directly perceived. Moreover, it would explain the sense in which values and meanings are external to the perceiver. The affordances of the environment are what it offers the animal, what it provides or furnishes, either for good or ill.The verb to afford is found in the dictionary, but the noun affordance is not. I have made it up. I mean by it something that refers to both the environment and the animal in a way that no existing term does. It implies the complementarity of the animal and the environment."*[4]

Why is this way of thinking relevant to the perception of tectonics?

*"When architects design the buildings of the future, they face a number of complicated challenges, due to a more ecological understanding of architecture. The question is how all these challenges will affect the way we look upon the relation between the parts and wholes of the building structure. Are we facing a new understanding of aesthetics within a tectonic way of thinking and designing, which challenges the traditional concept of coherence?"*

When we look at the scope of this concept, its focus is on the direct perception of an object. Here the meaning of the things surrounding us is embedded neither in the thing itself nor in the memory and experience of the person perceiving them. The meaning arises from the meeting of the two, between the surrounding environment and us. The buildings we encounter offer a set of affordances and possibilities which we can interpret depending on our background and the specific situation we are in, the moment the encounter takes place. The affordance of an object is permanent (*"it offers what it offers – because it is what it is"*[5]), but the way we interpret it will vary from person to person and from situation to situation. Linking this to a tectonic approach means that the tectonic appearance of a building is permanent, since it points to the logic of the structure itself. But the way this logic is construed by the user can differ, since it is embedded in a concrete situation and set of circumstances. According to this aspect, there is no pure or true understanding of the tectonic expression of a building; the same structure can offer a thousand readings. This dynamic way of looking at perception calls for strong tectonic strategies, which can ensure that the logic of a given building structure can be embedded in cultural and social systems in constant transition.

**Challenging the Concept of Coherence**

When architects design the buildings of the future, they face a number of complicated challenges, due to a more ecological understanding of architecture. The question is how all these challenges will affect the way we look upon the relation between the parts and wholes of the building structure. Are we facing a new understanding of aesthetics within a tectonic way of thinking and designing, which challenges the traditional concept of coherence?

The next two paragraphs will present the studies of two very different houses (Can Lis and Loblolly House) in an attempt to discuss this question. Both projects are very strong tectonic statements, but they represent disparate approaches to the construction of an entity. Both building structures are complex, but the way they achieve this complexity and the way they appear are highly diverse: each represents a distinctive tectonic strategy.

**A study of Can Lis**

In 1971 the Danish architect Jørn Utzon started the construction of his own house, Can Lis in Porto Petro, a small village on the southeastern coast of Majorca. The house is situated on a long, narrow plot, a steep cliff facing the sea. The main concept of the house resembles the principles of Utzon's work on the layout of the unrealised project for his own house in Bayview, north of Sydney (1963-1965). This simple but powerful concept, establishing each function in an autonomous building unit, also forms the basis of the drawings Utzon made simultaneously with the development of Can Lis for a house in the Majorcan interior close to the village of S'Horta. The concepts of the two Majorcan projects were identical, but the second one (Can Feliz) was not completed until 1991-1994 in a much-modified version.[6]

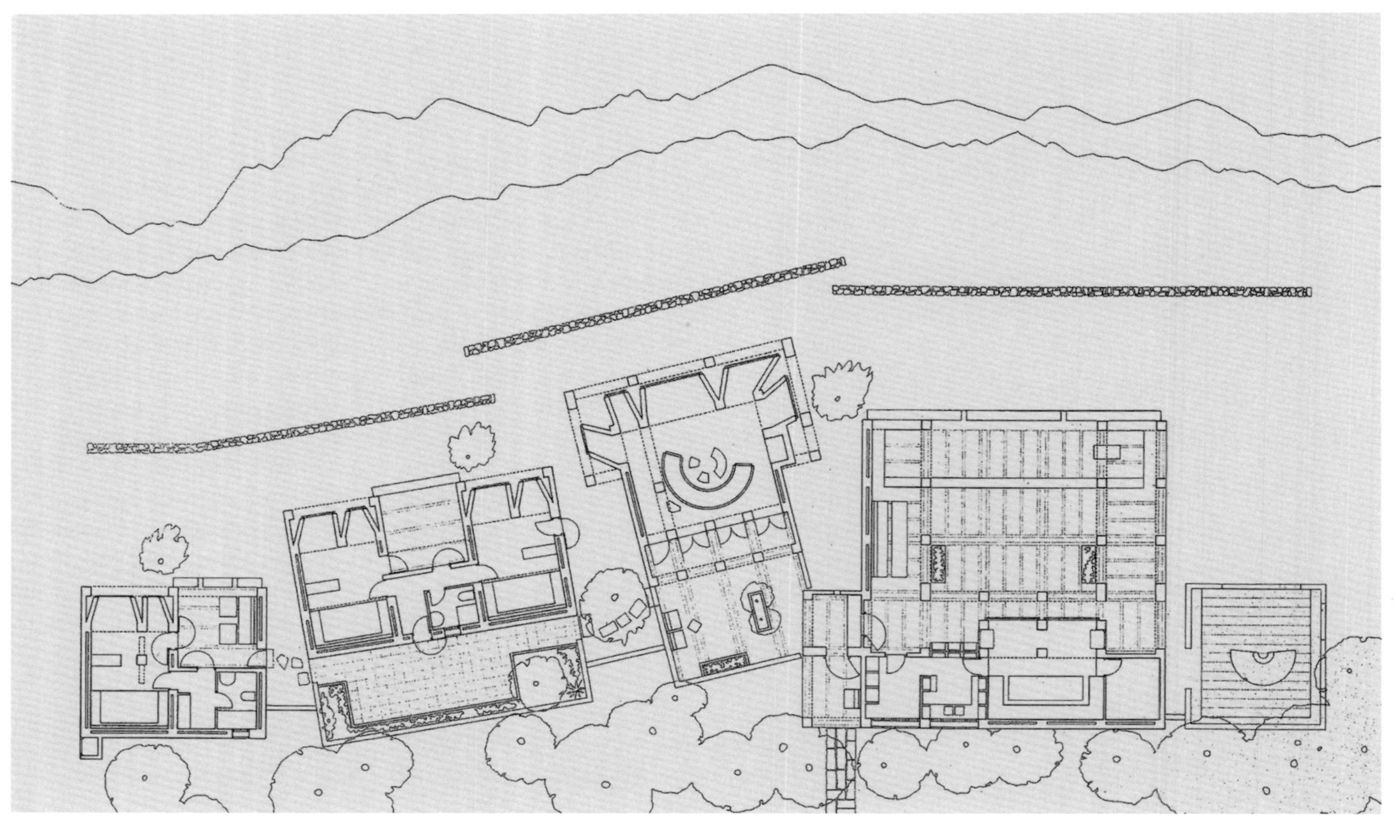

Can Lis – Utzon's original plan 1:200.
© The Utzon Foundation

Can Lis – view of the patio in front of the kitchen and dining area with the living-room building in the background. The buildings almost merge with the surrounding natural landscape.
© Ulrik Stylsvig Madsen

Can Lis is composed of four individual units, aligned on the narrow site, all facing the sea (plus one more unit, which only functions as an outdoor space). The units house different functions: kitchen/dining, living room, bedrooms and guest room. In close connection to each of the units is an outdoor space, reflecting the function of each of the small buildings. There is a clear boundary between the buildings, the walled outdoor spaces and the surrounding natural landscape of the site. When you move from one unit to the other, you go through a narrow passage where nature is wedged in-between the buildings (as seen on ill. 1). The building complex can be seen as small, cultivated enclaves, placed in the raw landscape and following the contour of the coastline. This concept of working with platforms and plateaus in contrast to the surrounding environment can be seen in many of Utzon's projects. He describes the origin of this concept in the text *Platforms and plateaus*[7] (1961), in order to set the scene for the layout of the Sydney Opera House.

Can Lis is constructed from the same stone as the rock on which it is built. In this way, the building almost becomes a continuation of the surrounding landscape. Building with stone blocks is part of the traditional building practice on Majorca. Since the stone material is very porous, it would normally be plastered to protect it from humidity. In Can Lis, the stone blocks remain raw and unprotected with clear marks from the circular saw used to cut the material. In order to prevent moisture from penetrating the walls, Utzon developed a cavity wall system based on Danish building practice.[8] This innovation of the system allows the structure of the stone walls to be visible, thus creating a strong tectonic appearance.

### Constructing Complex Simplicity

It took about two years for the local craftsmen to construct Can Lis, a rather long building period for a house of this size. One of the reasons for this was the fact that Utzon developed many of the construction details as the building went along.This way of working made it possible to create an almost countless number of site-specific solutions and details, all based on the same very simple principle. Utzon visited the construction site on a regular basis, making changes and solving problems as they cropped up.[9]

The entire construction followed the same simple principles. The stone columns are built of 20 x 40 x 40-cm blocks. The solid walls consist of larger blocks: 40 x 40 x 40 or 40 x 40 x 80. Between the columns and over doors standard concrete I-beams support the roof structure. Concrete trusses with terracotta shells between them support the roof. The roof itself is flat with a minimum of slope.

This way of working on the construction reflects the additive principles on which Utzon focused at this stage in his career. He described these principles in the text *Additive Architecture*,[10] a year before starting the construction of Can Lis. But this way of thinking was not new to him. It was already present in the following quotation from the text *The Innermost Being of Architecture*, written in 1948.

Can Lis – looking at the ceiling of Lis' bedroom, lying in the alcove. © Ulrik Stylsvig Madsen

Can Lis – view of the patio towards the dining room and kitchen. The simple structural principle can be seen in the stone columns, the concrete beams and the smaller transverse concrete beams holding the terracotta shells (here painted white). © Ulrik Stylsvig Madsen

*"The true innermost being of architecture can be compared with that of nature's seed, and something of the inevitability of nature's principle of growth ought to be a fundamental concept in architecture. If we think of the seeds that turn into plants and trees, everything within the same genus would develop in the same way if the growth potentials were not so different and if each growth possessed within itself the ability to develop without compromise. On account of differing conditions, similar seeds turn into widely different organisms."*[11]

This understanding of how the same principle can accommodate different contexts and take on different forms is the key to interpreting the unique character of Can Lis. If we look at the detailing of one of the bedrooms, we see how variations in the finish of the surfaces create different atmospheres – from the raw, cave-like feeling in the main room to the more intimate, refined plastered and painted surfaces of the alcove. This illustrates how Utzon worked by focusing on the interface of a simple constructive principle and a deep understanding of everyday life.

By means of detailing built on simplicity, Utzon managed to construct a rich and complex building structure based on deep respect for the cultural, topographic and technological aspects of the context. In other words, one can describe the tectonic concept as constructing complex simplicity.

### The Building Structure as an Essence

In order to describe and summarise the specific character of the work of Utzon, I will introduce the theory of assemblage developed by Manuel DeLanda.[12] In the book *A New Philosophy of Society*, DeLanda describes the movement from essences to assemblages in the way we structure our society. Transferring this way of thinking to the way we design and construct buildings, it offers an understanding of the relation between parts and wholes. Utzon's specific approach could, in this way, be described as constructing an essence or a totality. DeLanda defines the approach this way: *"Wholes in which parts are linked by relations of interiority (that is, relations which constitute the very identity of the parts)."*[13] Based on traditional craftsmanship, Utzon constructed Can Lis as a seamless whole. All parts derive from the same principle; they are variations on the same theme. The basic principle of the construction is very simple and can be construed by the user directly, without the need for background information. The tectonic appearance of the structure points towards itself. The meaning of the building derives from the logic of its construction.

### A study of Loblolly House

Loblolly House was completed on Taylors Island, Maryland, in 2006. The house serves as the holiday home of the Kieran family. The owner, Stephen Kieran, is, together with James Timberlake, a founding partner of the architectural practice KieranTimberlake Associates, located in Philadelphia. In the design of the Loblolly House, the architects explored new ways of structuring the building process, based on their ideas of off-site production,

Loblolly House – the west façade seen from the shore of Chesapeake Bay. © Ulrik Stylsvig Madsen

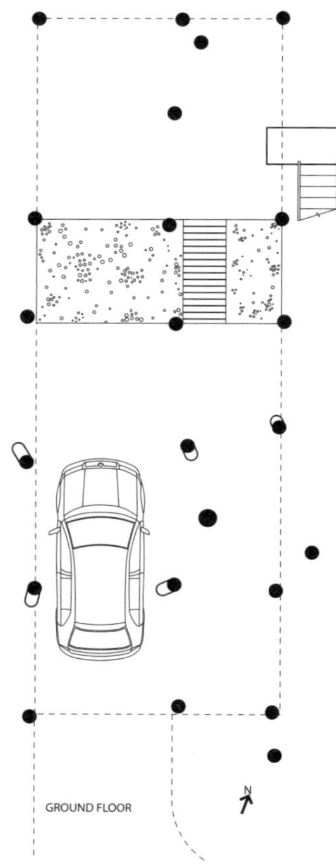


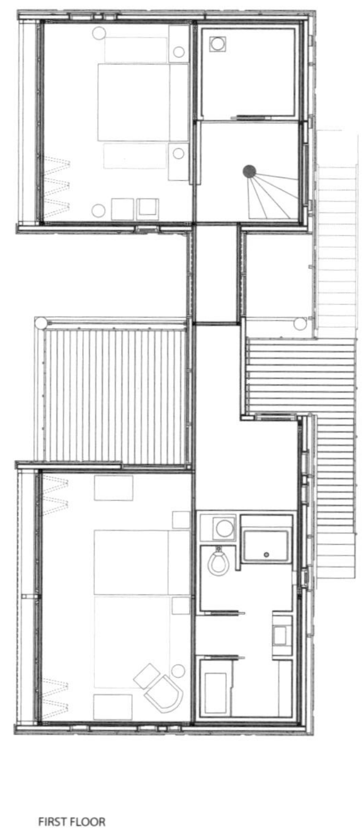


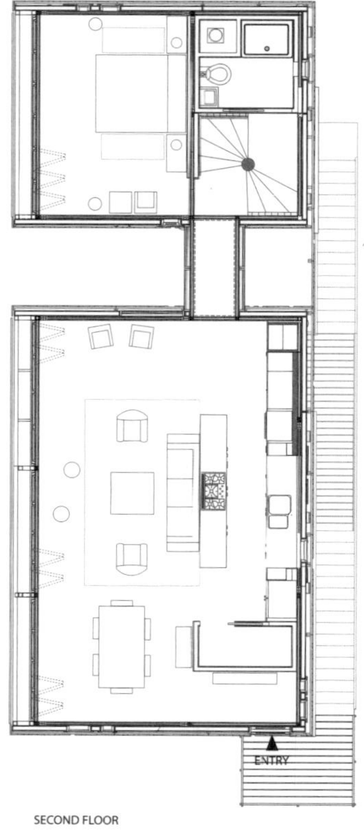


Loblolly House – plans 1:200.
© KieranTimberlake

which were presented for the first time in the book *Refabricating Architecture – How Manufacturing Methodologies Are Poised to Transform Building Construction.*[14] Building according to the concepts discussed in the book, the practice of KieranTimberlake has developed over several projects. The principles used in the construction of Loblolly House also formed the basis for the housing concept "Cellophane House", designed for the exhibition *Home Delivery* at MoMA, New York, in 2008.

Loblolly House is located within a forest of loblolly pines, facing the shore of Chesapeake Bay. The concept of the house is very simple. It is elevated one storey above the ground on wooden pillars. Beneath the house there is room for storage and parking. There are no outdoor spaces connected to the house, since it is set in the middle of the forest. The house itself consists of two separate, two-storey units, connected by a gangway covered in glass. The smaller of the units contains a guest room on each of the two floors, while the larger unit contains a master bedroom on the first floor and a living / dining room on the top floor. An outdoor flight of stairs on the east façade provides access to both floors from ground level.

The exterior of the house reflects the surrounding environment. Three of the façades (north, east and west) face the forest and are clad with a vertical sheathing of loblolly pine. In the design of the cladding, the architects were inspired by the flickering character of light coming through the leaves of the trees. The appearance of the west façade forms a contrast to the rest of the exterior with its big glass folding and hangar doors. This façade is designed to provide the maximum view and connection to the bay from all rooms within the house. A system of glass folding doors covers the entire façade, giving free access to the view. All doors can be slid aside, opening up the whole façade, and thus breaking down the boundary between inside and outside. In front of the sliding doors, a system of hangar doors serves as shutters which protect the façade from weathering. When open, they shelter the interior from the sun. The dynamic structure of the west façade makes it a tectonic tool for interacting with the surrounding nature, offering both shelter and an intense presence.

Loblolly House – the glass folding doors and hangar doors on the west façade of the house. © Kasper Sánchez Vibæk

Loblolly House – the meeting of the aluminium scaffold and the different kinds of floor and wall cartridges in the house. © Kasper Sánchez Vibæk

### Constructing Simple Complexity

The Loblolly House was built out of a number of larger elements produced off-site and then assembled on-site within a period of only six weeks. The design concept divides the elements into five categories: the piles (wooden columns forming the base of the structure by lifting the house one floor above the ground); the scaffold (an aluminium frame structure made of standard profiles with simple connectors); the cartridge (larger prefabricated floor, wall and cladding elements with integrated heating and electrical systems resting on the aluminium scaffold); the block (prefabricated bathroom and mechanical units placed on the floor cartridges); and the FFE (the fixtures such as kitchen cabinets and stairs, the movable elements such as the glass walls, and finally the furniture of the house).[15]

KieranTimberlake has described the principle of working with the building structure as a set of larger autonomous elements depending on each other in this way:

*"Quilting, not weaving. Dimensional certainty is a direct product of the parametric model. Each element is assembled into a geometric whole that must be 'closed in all directions'. That is, each element links to the next until they form a cohesive whole. [...] In current practice, a building is woven together as a series of systems. We measure as we go, ensuring the fit of each successive element. In contrast, the parametric model embeds geometric and dimensional certainty within it, as opposed to unearthing these details during construction. Through the agency of this digital tool, we become confident quilters, rather than tentative weavers."*[16]

Using the digital tool of BIM-modelling, one can (according to the quotation) control the building process in a new way, working with almost no allowance. Each of the larger elements within the construction can be produced individually, as long as one is able to control the interface between them. This way of structuring the building process became the core of the design concept and the appearance of the Loblolly House.

If we look at the meeting between the different elements of the construction (scaffold, cartridges and FFE), a clear tectonic statement is revealed. Each of the elements appears as an autonomous system, underlining the contrast between them (this can be seen on Ill. 9 showing the meeting of the aluminium scaffold and the floor and wall cartridges). Although the appearance of each of the individual elements is very simple, the appearance of the construction as a whole becomes very complex. In other words, one can describe the tectonic concept as constructing simple complexity.

### The Building Structure as an Assemblage

To summarise the study of Loblolly House and the way the project reflects a certain approach to the relation between parts and wholes, I will look at the work of Manuel DeLanda again. He defines the concept of assemblages in this way:

*"Assemblages are made up of parts which are self-subsistent and articulated by relations of exteriority, so that a part may be detached and made a component of another assemblage."*[17]

When constructing the Loblolly House, KieranTimberlake used the concept of "quilting" to assemble the large, industrially produced elements into a whole: not a seamless whole like Can Lis, but more like an area of tension. The meaning of the construction arises from the conflict between the different tectonic approaches of the parts. Every element or detail is in itself very simple and logical (clear tectonic statements), but the logic of the overall structure can be difficult to read. Being an assemblage, the construction points towards an understanding beyond the structure itself.

**Constructing Immediacy – a Tectonic Strategy**

Based on the studies of Can Lis and Loblolly House, an answer to the question raised in the beginning of this text starts to emerge. If one is to meet the challenges of adapting the way we build to ecological principles and new, industrialised ways of production, we need to change the tectonic strategies used in the design of our buildings. One of the lessons learned from the two cases is the importance of working with clear and simple tectonic concepts, which can be understood by the user directly, without the need for deep knowledge of a specific field. This ensures that the logic of the construction can be construed immediately. Constructing immediacy as the core element, the way one designs buildings can be a way of ensuring that the logic of a given building structure can be embedded in cultural and social systems, which are in a state of constant transition. Using this as a tectonic strategy will ensure that the potential of the construction can be recognised by the user, and the building will be an active part of his or her everyday life. But focusing on the immediacy of the tectonic structure is not enough to meet the challenges of the future. One also needs to redefine the understanding of the relation between parts and wholes!

**From Essences to Assemblages – the Strategy of the Future?**

Looking at the relationship between parts and wholes within the building structure, could the concept of assemblages be the answer to the challenges one faces? Based on the discussion in this text, it seems like a fruitful strategy. Instead of creating seamless wholes, focusing on the individuality and independency of the different parts of the construction would open up new possibilities. The complexity of the challenges calls for tectonic strategies that can embrace a broad spectrum of solutions embedded in very different fields of knowledge.

But will this lead to the death of coherence: a new design paradigm, where anything goes?

The answer to this is "no". Entities are still being created within the concept of assemblages. But the creation of coherence will have to follow new rules. The building structure will no longer be a homogeneous whole, but rather a heterogeneous collection of parts. This calls for a tectonic strategy which focuses on the ability of the parts to intensify and clarify the differences between them. It might not be possible to perceive a clear logic in the overall structure. But the immediacy of a tectonic approach in the different parts of the construction and the dependencies between them can be construed. This way of thinking opens a more dynamic way of structuring the building and thus a possible solution to the challenges of the future. But it calls for dramatic changes in the way architects think and the way they design buildings. New aesthetic approaches and tectonic strategies are going to be necessary. The question is,

*are we ready for this change of paradigm in order to ensure the well-being of our planet as a whole – or do we really have a choice?*

**Ulrik Stylsvig Madsen**
is an associate professor at the Royal Danish Academy of Fine Arts, School of Architecture. In 2009 he defended his PhD thesis, which focused on architecture as an important part of the creation of identity, both for organisations and individuals. Since then his research and teaching has focused on tectonics, architectural quality and ecology. He is the author of several articles on these subjects and the co-author of the book *Building the Future – visions within industrialised housing 1970-2011* published in 2012.

Loblolly House – the east façade with the wooden cladding, leaving openings for us to see the wall cartridges underneath. The stairways forming the entrance to the building are also visible. © Ulrik Stylsvig Madsen

1 __ Marco Frascari, "The Tell-the-Tale Detail," *VIA7 - The Building of Arcitecture* (1984), p. 23.

2 __ Anne Beim, *Tectonic Visions in Architecture* (Copenhagen: Kunstakademiets Arkitektskoles Forlag, 2004), p. 53.

3 __ Beim,*Tectonic Visions in Architecture, p.* 52.

4 __ James J. Gibson, *The Ecological Approach to Visual Perception* (Hilldal: Lawrence Erlbaum Associates, 1986), p. 127.

5 __ Gibson, *The Ecological Approach to Visual Perception*, p. 139.

6 __ The drawings for the 3 projects can be found in the Utzon Archives. I have been studying copies of the material in the book by Jamie J. Ferrer Forés, *Jørn Utzon. Works and Projects* (Barcelona: Editorial Gustavo Gili, 2006), pp. 230-231, 270-275 and 292-297.

7 __ Jørn Utzon, "Platforms and plateaus: ideas of a Danish architect," *Zodiac no. 10* (1962), pp. 112-117.

8 __ This cavity wall is described by Kim Utzon (the son of Jørn Utzon) in the text: Kim Utzon, "Eget feriehus på Mallorca, 1971-," in *Arkitekternes egne boliger- med arkitekternes egne ord*, ed. Mogens Stærk, Mogens and Kirsten Grønborg (Århus: Arkitektskolens Forlag, 2010), p. 404.

9 __ The building process has been described in the following text by Tobias Faber: Tobias Faber, "To huse på en Midelhavsø," in *Utzons egne huse* (København: Arkitektens Forlag, 2004), p. 87.

10 __ Jørn Utzon, "Additive Architecture," *Arkitektur no. 1* (1970), p. 1.

11 __ Taken from an English translation of the text "Arkitekturen Væsen" published in *Grønningen* (exhibition catalog) in 1948: Jørn Utzon, "The Innermost Being of Architecture," in *Utzon, Inspiration, vision, architecture*, ed. Richard Weston (Hellerup: Edition Bløndal, 2002), p. 10.

12 __ Manuel DeLanda built his theory on the concept of assemblage developed by the French philosophers Gilles Deleuze and Félix Guattari.

13 __ Manuel DeLanda, *A New Philosphy of Society* (London: Continuum, 2006), p. 18.

14 __ Stephen Kieran and James Timberlake, *Refabricating Architecture – How Manufacturing Methodologies Are Poised to Transform Building Construction* (New York: McGraw-Hill, 2004).

15 __ The structure of the house has been described in the book by Kieran and Timberlake, *Loblolly House*, pp. 43-139.

16 __ Kieran and Timberlake, *Loblolly House*, pp. 100-101.

17 __ DeLanda, *A New Philosphy of Society*, p. 18.

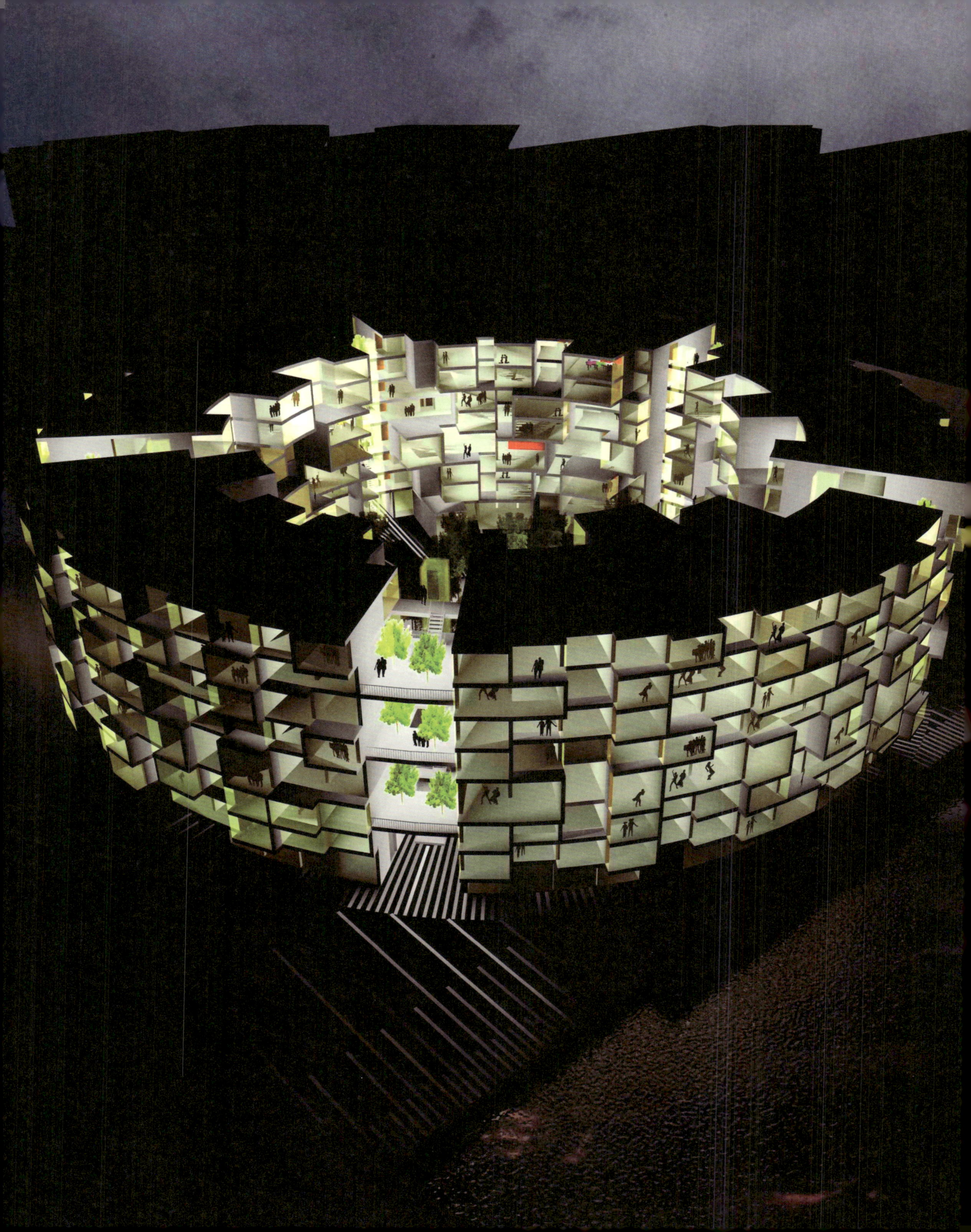

Peter Thorsen

# Practice as a Constructional Craft!

Lundgaard & Tranberg Architects are a partnership which is based on a common understanding of architectural ideals and visions, developed during more than 25 years of practice. Among the core values of this consensus are the following:

First: we are unconditionally *regional and contextual* in our professional understanding, in contrast to the rather universal approach to architecture that currently seems to predominate. Accordingly, we find that every task should be addressed with an explorative curiosity, looking for the specific features of the location and the programme in order to lend the scheme an original, unmistakable identity and root it firmly in its surroundings. This requires a physical presence and personal identification, which is why we consider our geographic field of interest to be primarily Denmark and the other Nordic countries. We have no stated ambition or strategy to work on the global market on a large scale.

Second: architecture deals with real people. It is a kind of applied sociology, and we think it should be considered a public issue. It affects and involves not only clients and end users in a narrow sense, but basically the whole of society as well: you and me, us all, everybody. Consequently, in our view, architecture should be generous, adding ambitions to the basic programme. It should be inclusive and considerate and, above all, human. Beyond the strict fulfilment of programmatic requirements, it should strive to generate activity and stimulate social interaction; appeal both to users and those just passing by; affect our senses and tell stories; link past with future; and age with dignity, surprise and delight.

Third: it is not architectural form in itself we find interesting, but rather the inherent relationship between form, content, construction and materials, and the possible integration of these into a seamless whole, turning every individual project into a resolved architectural statement. In other words, we consider architecture to be a *constructional craft* which requires considerable knowledge of building technology and processes in order to achieve tectonic clarity in individual designs, and to contribute to innovation and the further development

Tietgenkollgiet, competition scheme, 2002. Visualization © LT Architects of architecture in general. In our understanding, genuine design quality is a precondition for permanence in architecture and thus the most meaningful contribution towards sustainable architecture, ensuring that buildings, cities and landscapes remain useful and attractive for generations.

**Tietgenkollegiet – Exploring the Tectonic Dimension of Architecture**

Of all our projects, to my mind Tietgenkollegiet is the one which, in the most consistent and clarified manner, demonstrates a tectonic approach to architecture. The project has its origins in a competition launched in 2002 by a foundation associated with one of Scandinavia's largest financial institutions. A considerable shortage of student housing in central Copenhagen at the time was a major motivation for this initiative, as was the fact that the *kollegium* (Eng. hall of residence, Amer. dormitory), as such, seemed to have become much less attractive and popular at the turn of the millennium than before, particularly in comparison with the peak period around the 1970s. The latter was a major concern, since this housing type has been a common tradition in Denmark for centuries, with the purpose of providing affordable accommodation for students, based on a combination of individual privacy and communality, usually in the form of shared household facilities.

At the same time the competition brief was unusually open and visionary, calling for an original architectural idea that could eventually be developed and realised as *the kollegium of the new millennium*, a state-of-the-art example of architecture which would revitalise and renew the attraction of this special type of dwelling, also representing a reference for the future development of student housing, even at the international level. The client's vision was to create a student housing environment "...that will inspire a sense of community and stimulate interdisciplinary, intercultural encounters and a tolerant atmosphere ..."

**Concept**

The idea of a circular building cropped up early on in the design process and seemed to be intuitively very appropriate and logical in terms of both the urban context and the internal organisation of the building. The site is located in Ørestad, a recent, large-scale urban development in close proximity to downtown Copenhagen. The site is adjacent to Copenhagen University and is flanked by two canals that constitute primary elements in the Ørestad master plan: one in a serpentine flow, the other linear in form. To the southwest, the site connects to a vast meadow, a nature reserve, which extends for several kilometres and provides unique recreational and landscape potential, given its central placement in the city. The concept can be seen as a critical comment on the way Ørestad was planned and developed, based on rigid, uniformly proportioned and oriented block structures, resulting in an uncertain urban definition and architectural identity. We could see an opportunity to introduce a more compact and nondirectional design, which would open up the neighbourhood and connect better with the open landscape, while adding significant architectural energy and personality to the location.

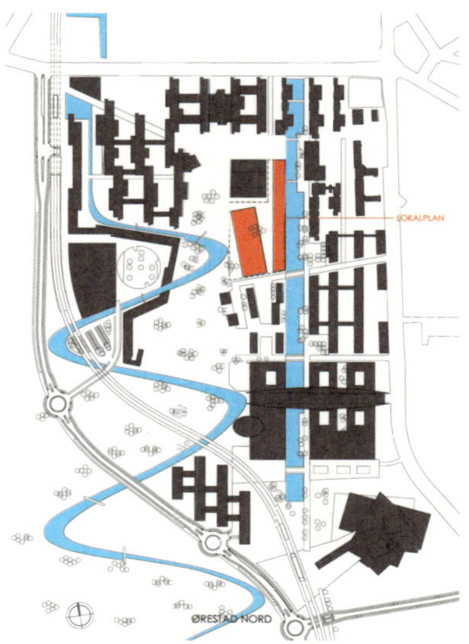


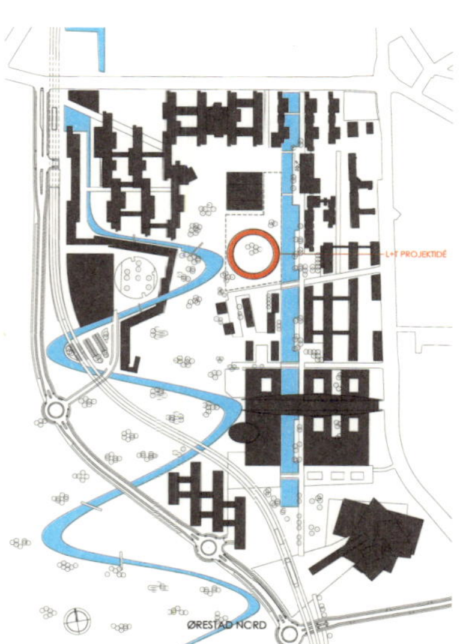


Aerial view of central Copenhagen and Ørestad. © LT Architects

Ørestad, masterplan scheme. © LT Architects

Ørestad, competition scheme. © LT Architects

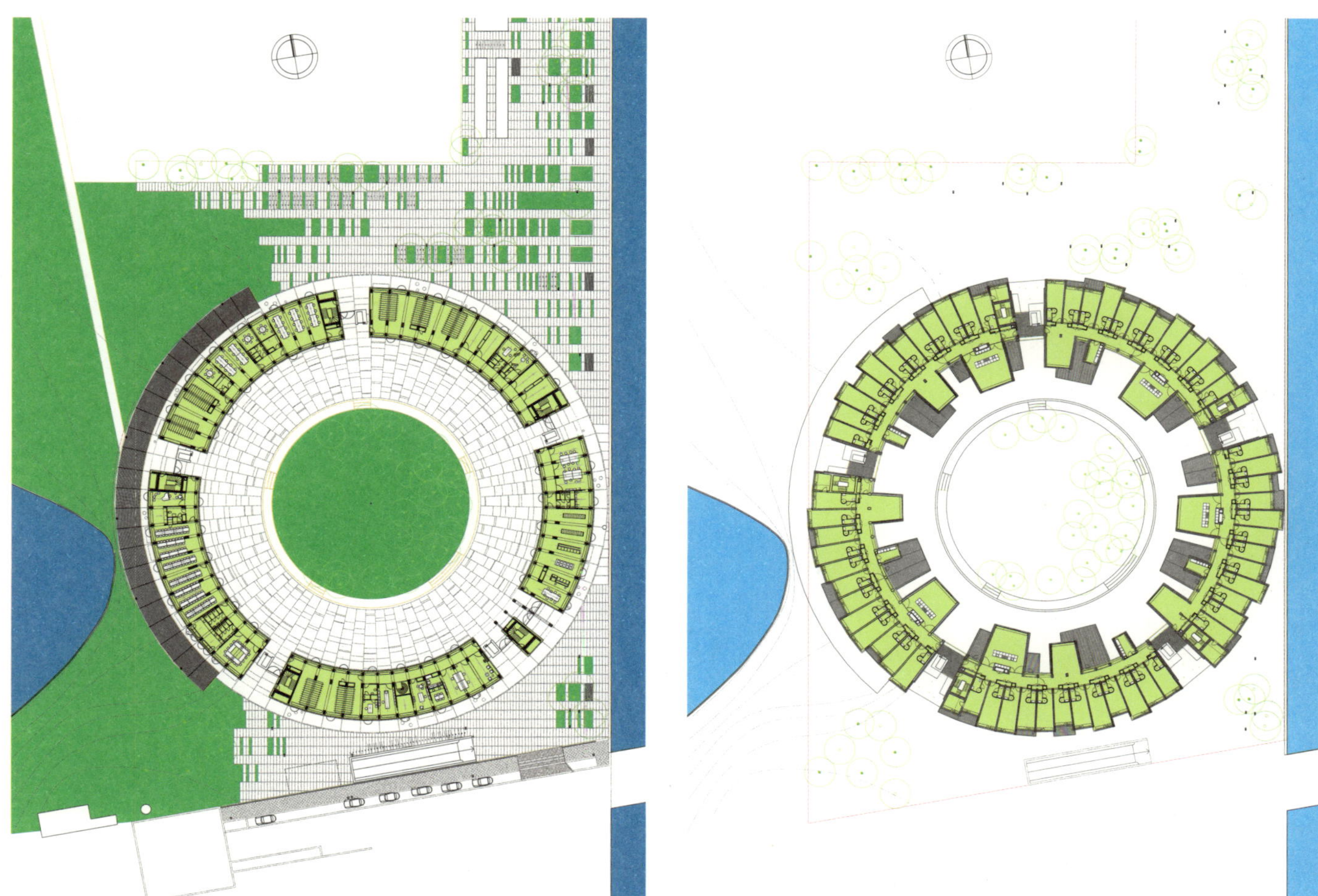

Plan of ground floor, final project. © LT Architects
Plan of residential floor (4th floor), final project. © LT Architects

However, the proposed cylindrical form did not comply with the rectilinear development plan of Ørestad, and so, for some time during the design process, we put the idea aside and investigated alternative solutions. However, none of these had the strength and clarity of the circular scheme, which meant that discussions kept returning to the original concept, as if it had a strong, insistent will of its own. Most architects have probably faced this dilemma, not least when developing competition proposals:

- Should one be loyal to one's intuition and challenge the programmatic constraints, risking disqualification and a painful loss of resources?

- Or does one "play it safe" and provide a response that accommodates exactly what is asked for?

In our experience, there is no evidence that one approach to this dilemma has a better success rate than the other, and we eventually decided to stick to the circular scheme.

Exterior, competition concept, 2002.
Exterior, as-built, 2006
Rendering and photo LT Architects

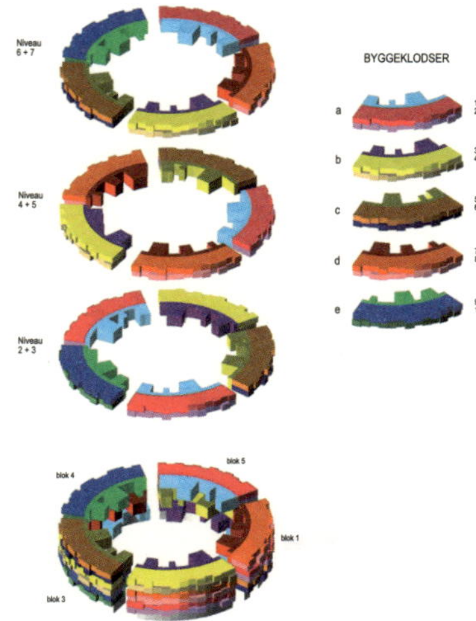


Morphological structure.
© LT Architects

### Design

The basic form determines the internal organisation of the building. All individual student rooms are placed to the outside of the cylinder with views of the surrounding cityscape, and the communal spaces to the inside, facing and overlooking the central courtyard. The ground floor consists of communal facilities for the entire kollegium, including administration, study and meeting rooms, laundry, workshops, lounge and assembly hall (party room), as well as indoor bicycle parking. The upper six floors house the building's 360 private units, organised in 30 groups, each with 12 individual rooms and 3 communal spaces: kitchen/dining room, lounge and pantry. The building is subdivided into five segments by open portals, which provide access from the surroundings and internally connect the individual floors with stairs and elevators. On all floors horizontal circulation is arranged at the inner perimeter, linking the portals and offering sequences of framed views of the courtyard.

One particular design challenge was to downscale and subdue the formalistic, monumental character of the cylindrical form, which one does not immediately associate with a vibrant housing environment for students. This prompted us to develop variations on the basic circular theme and to incorporate a certain elasticity and dynamic in the design: to emphasise the individual as opposed to the communal, and to give every location its own significance within the overall context. This resulted in the characteristic, crystalline expression of the building's exterior. A variation in size of the private units, as suggested in the competition brief, provided the tool for developing the outer façade into an apparently random composition of alternating projections. On the inner façade, we envisioned that the communal spaces of the various dwelling groups would be expressively cantilevered volumes, pointing into the courtyard and to the imaginary centre of the overall composition in a similar, arbitrary configuration.

Naturally, we investigated the structural viability of this concept closely with our engineers during the competition phase. Apart from this, though, the entry is probably the most conceptual we have ever submitted for a competition, leaving room for further project development in close cooperation with the client, as intended in the brief. Main goals for further planning, following the competition, included simplification and the provision of maximum scope for repetition, in order to further clarify and refine the complexity of the design, to optimise constructional feasibility and minimise production costs. The expressive variations of the design were morphologically simplified and systematised, resulting in a compositional order that has its own logic both geometrically and structurally, though this is (intentionally) not legible in the completed building.

### Structure, materiality, textures

To be precise, Tietgenkollegiet is *not* a round building. It is a 75-sided polygon, based on an evenly distributed, concentric modular grid which reflects both the spatial and the structural organisation of the building. Only

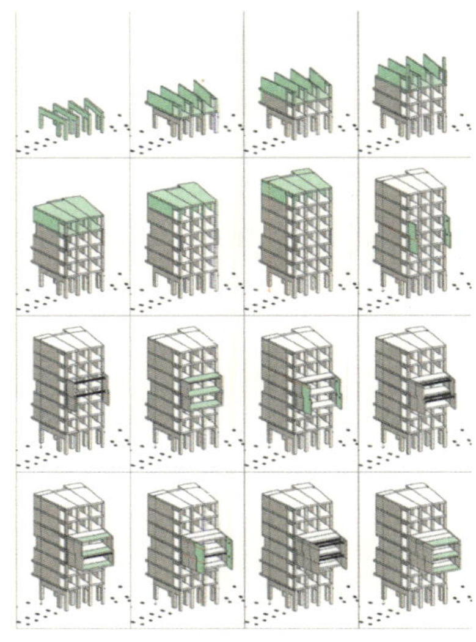

Structural concept and assembly sequence. © LT Architects

the basement, containing underground parking, various technical facilities and storage spaces, together with the underlying foundation and piling structure, is perfectly circular. The ground floor consists of radially oriented, site-cast, 3-legged concrete frames, which provide the necessary load-bearing capacity (including the loads of the cantilevered structures above) and still leave a flexible layout and open use of the ground floor spaces. We consider the latter relevant in terms of meeting future changes in needs and behaviour patterns of the residents.

On the upper six floors, the main structure consists of two independent, interacting structural systems. The structural backbone, the main building, is a highly rational and cost-efficient structure of precast concrete walls, columns, lintels and slabs. The cantilevered volumes, containing the communal spaces of the dwelling groups, are "hung", projecting up to 8 metres out from the load-bearing structure of the main building, and, fully loaded, weighing around 250 tons each. To cope with this challenge we developed a structural principle inspired by the technology of bridge building. The side walls of each box consist of two-storey-high, prefabricated concrete elements, which are mounted on precast brackets fixed to the circular building's structure, and subsequently attached with horizontally mounted, high-tensile, post-stressed steel cables. In order to prevent the circular frame building from tipping over, caused by the load impact of the cantilevered volumes, the primary structure is vertically anchored to the basement and foundation structure by further post-stressed steel cables at the outer perimeter of the structure. Although a well-known building technology, its unorthodox and unprecedented use necessitated extensive planning and tests before assembly. The combination of prefabrication with on-site execution proved to be both cost- and time-efficient, not least due to a high rate of repetition and off-site prefabrication, and because assembly could be done using ordinary cranes, which completely eliminated the need for scaffolding.

After tender and contracting, construction costs proved to be almost 10 percent under budget target, providing a comfortable buffer for materials, design features, fittings, artwork, etc.

Since the overall form, spatial and functional organisation, and structural system are so closely interlinked, the concrete structure is consequently visible throughout the building. The details, materials and colouration are carefully harmonised in order to emphasise this further, balancing and contrasting the raw concrete with softer, more delicate textures in order to provide an unmistakably domestic atmosphere.

The façades are clad with panels of tombac, an alloy of copper and zinc, which weathers to a varied range of dark, matte, red-brown tones, similar to bronze. The material is very resistant and suitable for the Danish climate, and is furthermore easy to process and adapt to the complex two-dimensional geometry that characterises Tietgenkollegiet. The dark colouration of the façade is fully intentional, in order to concentrate visually and to "hold" the expressively "spinning" main form in place, and to let the building's eyes (the

Cantilevered structures, preparing for vertical steel cables. © LT Architects

Construction site, view from crane tower. © LT Architects

Façade cladding of tombac, right after assembly/before weathering.© LT Architects

Private unit, competition concept. © LT Architects
Private unit, as-built. © LT Architects

Graphic print on plywood panels along corridors. 'Unfolded' elevation of one building segment © LT Architects/ artwork and signage Aggebo & Henriksen

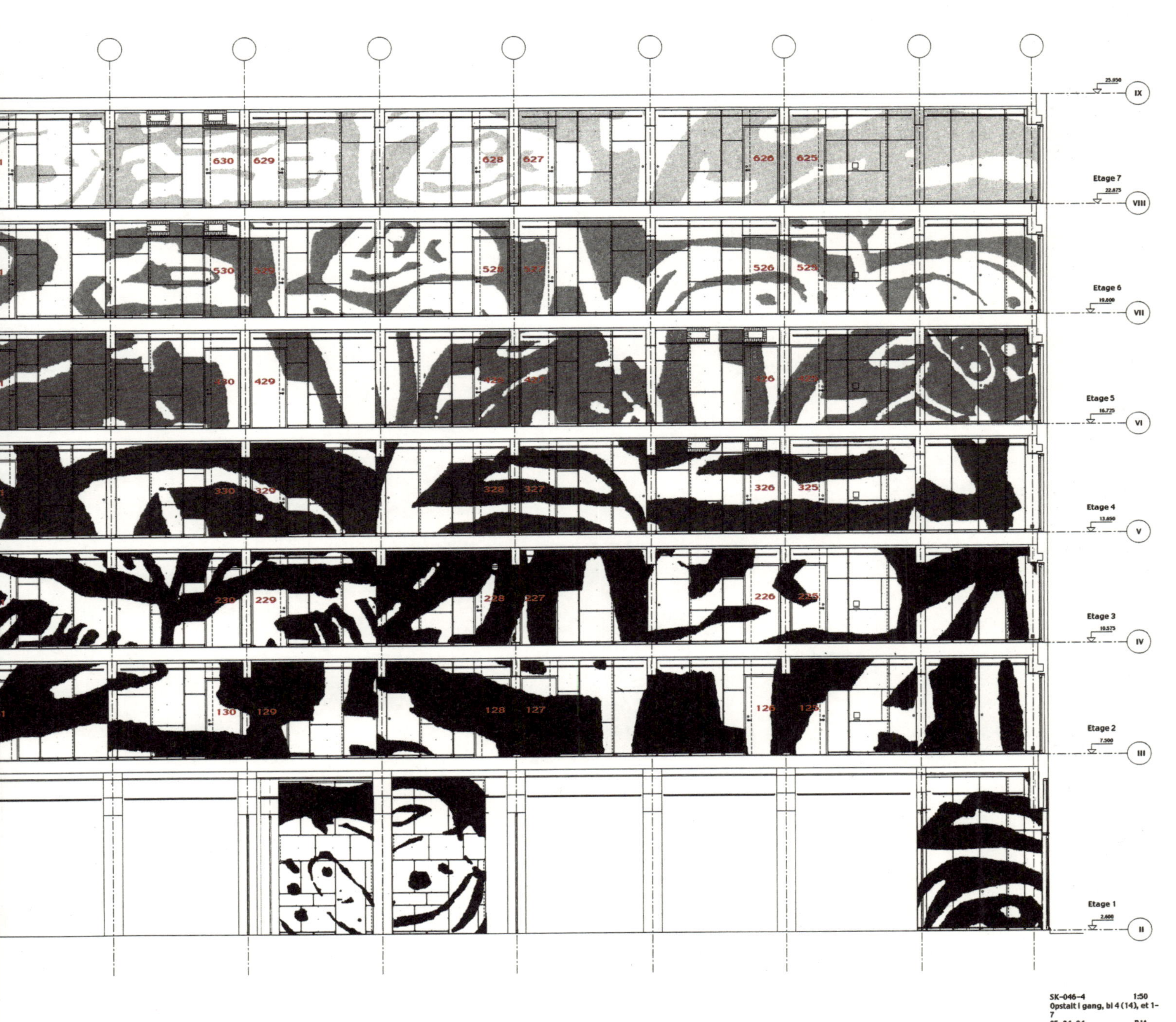


Next page: Night view of courtyard.

windows) light up and shine, enabling the vibrant life within the building to express itself. Glazing frames and sliding louvres are made of oiled oak, with the intention of adding a noble, domestic character to the otherwise rather industrial atmosphere of the building.

Flooring throughout the building has a screed of magnesite, an extremely durable and tonally rich material, previously known mainly from industrial facilities and warehouses, etc. Walls along the curved corridors, as well as the built-in furniture walls of the private units, are clad with panels of ordinary pine plywood. In contrast to the rough concrete, this provides a rather sensual and fragile texture, at the same time uncomplicated and inexpensive to replace, if necessary. A graphic print of organic images, in combination with the necessary signage, was applied to the panels along the corridors. This makes every panel unique and thus corresponds to the basic idea of the building and the core value of the kollegium: the importance of the individual and of singularity, based on an overall collective equality.

### Tectonics – the Synergy of Space, Structure and Materiality

In conclusion, Tietgenkollegiet represents a further development and clarification of what our practice has worked on for years: the synergy and interaction between space, structure and materiality, the tectonic dimension of architecture, in our understanding of the term.

The project has received a number of prestigious awards and prizes which have naturally been a wonderful recognition and appreciation of our practice. But even more rewarding and satisfactory has been discovering how the 360 young people who inhabit the kollegium have taken possession of it in a way that not only lives up to our visions and most optimistic expectations, but also goes well beyond and above them.

*In the end, architecture is all about people!*

Some comment that Tietgenkollegiet is "spectacular". This is the precise opposite of our intention. Indeed, it is something that in general we deeply oppose, as one of the major pitfalls of our profession today: the pictorialisation and vulgarisation of architecture, and lack of regard for the fundamental complexity and depth of architecture's role.

**Peter Thorsen**
is partner and managing director of the Danish architectural practice Lundgaard & Tranberg Architects. He was the project architect on the development of the *Tietgenkollegiet* project. During the last decades Lundgaard & Tranberg have gained international acclaim for their work. Looking at the distinctive character of their designs, one recognises a very strong focus on the tectonic dimension of architecture. The practice's most celebrated works include the Royal Danish Playhouse in Copenhagen, SEB Bank, Sorø Art Museum and Tiegtenkollegiet Dormitory. In 2012 the book *Tietgen Dormitory / An Imaginary Journey around a Real Building* was published, presenting the hall of residence and the working methods of the practice.

Community space (kitchen), as-built. © LT Architects

Ole Egholm Pedersen

# The Tectonic Complex of Concrete

In this article I propose a definition of tectonic concrete casting. This definition is subsequently tested in two design experiments, which try to exemplify how complex concrete elements could be produced in an era of mass customisation. While the experiments are not fully developed as building principles ready to use in construction, the aim is to exemplify how the understanding of "tectonics" and the attention to new manufacturing technology available in industry may in the future lead to concrete constructions that perform better both ecologically and architecturally.

Cultural and historical factors, personal preferences and the wish to build in accordance with the qualities of a given place have resulted in a backlash against commonly used, repetitive concrete forms. Whenever the brief and budget allow, architects try to apply a degree of individuality to concrete architecture. This is evident when standardised concrete elements are placed in varied formations to mask the appearance of repetition.

The backlash against repetition has also affected the attitude towards concrete as a material in a negative way. Concrete is rarely allowed to serve as the cladding of a building, even though it is one of the most durable and weatherproof materials available. Usually an actual concrete house is concealed behind an additional façade, which is applied in a different material. This increases the resources needed and the complexity of contemporary architecture. In short, the architectural potential, which lies in working actively with material characteristics and a structural reading embedded in architecture, is eliminated.

One could ask whether it is possible to identify concrete casting techniques in which there is a closer connection between the material, the technology and the form of the concrete element than is the case today, and whether we can address two problems in the current production of concrete elements:

ReVault pavilion. Detail of the structure © Niels Martin Larsen

- The lack of connection between the intention of creating variation in our built environment and the current ways of shaping concrete elements.

- The environmental challenges, which primarily call for a reduction of waste products generated in the production of concrete elements.

**Concrete Tectonics. Tectonics and Concrete Casting**

I use the term "tectonics" as a conceptual framework which seeks to define a close relationship between material, technique and the resultant form. This understanding of tectonics is elaborated in the article "Nature, Culture, Tectonics of Architecture" found elsewhere in this anthology. Dealing with material and technique as an axiom in architecture is useful in order to focus the study, emphasising the current technological and material situation.

When considering what tectonic concrete casting might be, one needs to contemplate the fact that a concrete casting is the imprint of a mould. This implies two processes: the concrete casting and subsequent assembly of the concrete structure; and the actual making of the mould. The first process concerns the making of geometric forms with concrete - in other words, the construction of the *building*, which involves a *material* (concrete) and a *technique* (casting). If the concrete is cast as elements, the technique also includes assembly. The making of the *mould is a* subset to this technique, which for its own part involves a *material* (the mould) that has a *form* subjected to a *technique*.

**Concrete Casting Relations**

This understanding of making built forms in concrete is hence referred to as *concrete tectonics*. The three elements of concrete tectonics may be expressed as follows:

- Material refers to the concrete itself: that is, stone, sand, aggregates, water, additives and possibly reinforcement.

- Technique refers to the casting process and the assembly process: the action of making the mould, casting the concrete, and post-processing the cured concrete into a construction (post-tensioning, assembly, etc.).

- Form refers to the geometric form of the cast concrete.

**Mould Making Relations**

The making of the mould is referred to as *mould tectonics*, because the *mould itself* may be viewed as a relation between material, technique and form. The three elements of mould tectonics may be expressed as follows:

- Material equals the mould material.

- Technique refers to the making of the mould: that is, the addition, subtraction or deformation of materials, which creates a negative volume in which to pour the concrete.

- Form refers to the geometric form of the mould.

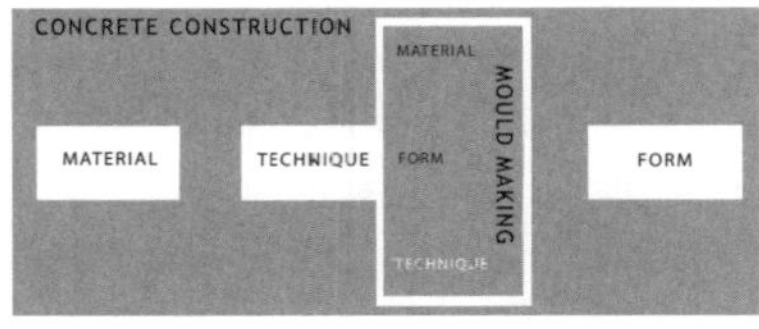


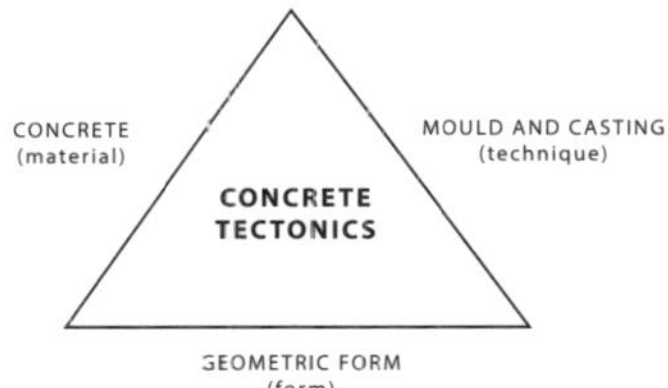


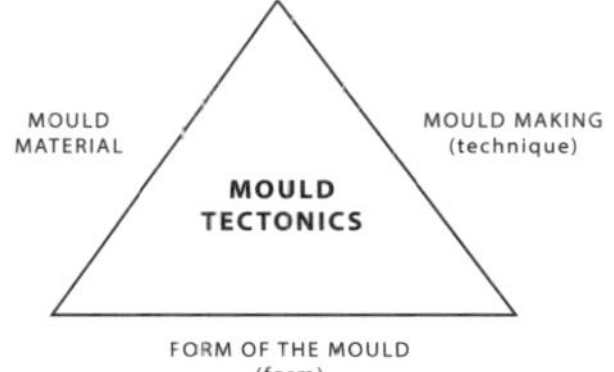


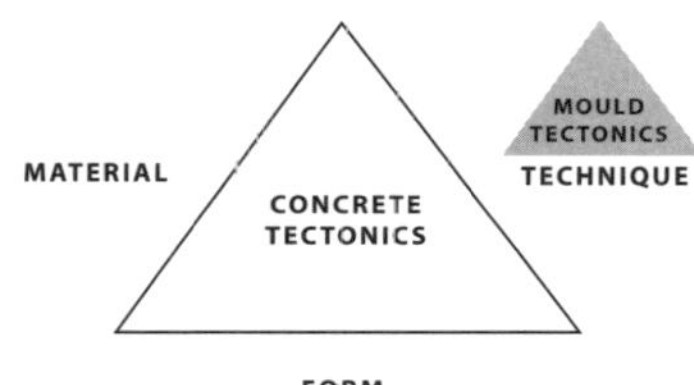


The mass-produced sandwich element is used extensively in contemporary architecture, giving buildings an appearance of repetition. © Ole Egholm Pedersen

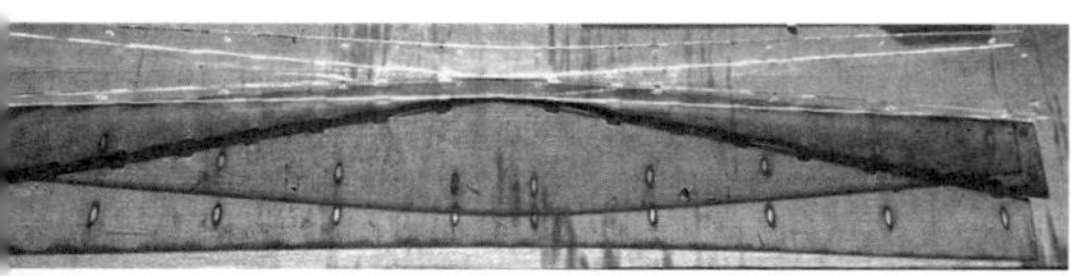

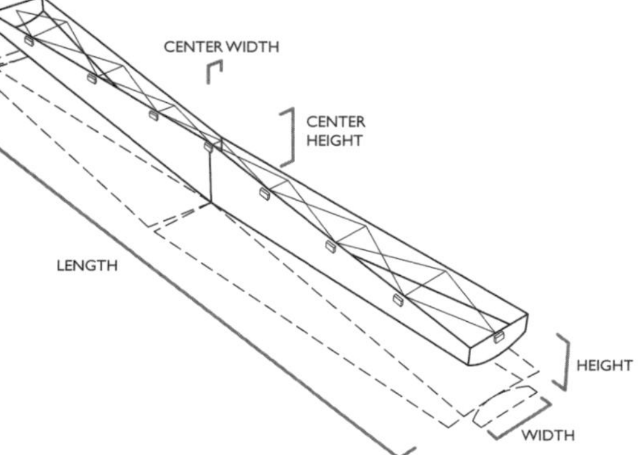


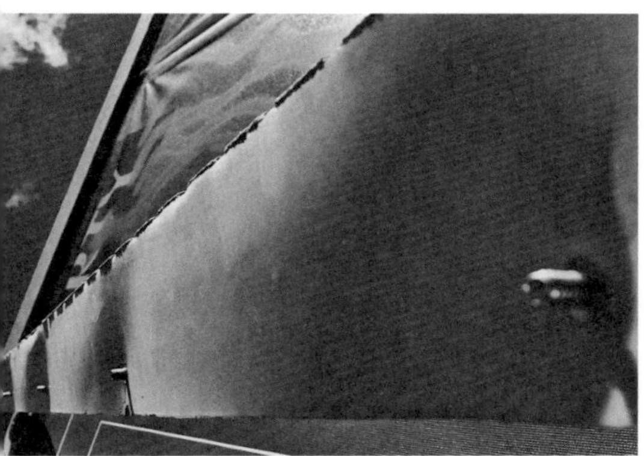

Folded PETG plastic. © Ole Egholm Pedersen
A parametrically defined casting mould. © Ole Egholm Pedersen
Cast prototype. © Ole Egholm Pedersen
Assembly of three columns and four beams. © Ole Egholm Pedersen

The two design experiments presented here focus on *mould tectonics*, since the forms in concrete rely on the geometry and materiality of the moulds in which they are cast.

**Tectonic Relations between Mould and Concrete**

The form of the mould is a complicated matter, because it is also part of the concrete casting technique, and the *form of the mould* should also be evaluated as part of the *technique* of concrete tectonics.

**Design experiment I: Fold**

In the design experiment "Fold" I utilise the conceptual framework described above to investigate how attention to a specific material and a technology with potential for mass customisation may yield complex concrete elements.[1]

The outset of the design experiment is the choice of a technology that is both fast and readily available in industry: laser cutting. Laser cutting can cut through flat sheets of recyclable material, such as plastic, at high speed.

The obvious technique suitable for translating flat sheets of material into three-dimensional casting moulds is bending or folding. In fact, this is the simplest way to create a three-dimensional volume from a two-dimensional surface. A folding test of various plastic types reveals PETG plastic as suitable: it is somewhat ductile, while other plastics are brittle and can snap. Thin sheets are foldable along a bevelled edge; thicker PETG can be bent, provided dashed fold lines have been made with the laser cutter.

In order to allow for the production of discrete concrete elements, I define a set of parameters that can be described digitally. The process involves generating a system that guides the placement of each concrete component in a 3-D model, then generating the component on the basis of the information given by the system, and finally unrolling the component geometry into a casting mould.

Having established this technique, I produced a full-scale structure of five columns and beams. The form of each individual component is a result of an optimisation of the component itself, and its placement in the overall system. The structure is an example of a prototypical, mass-customised, concrete element, column-beam construction, in which each element is unique.

Based on the experiences from the first design experiment, I developed a new design experiment in collaboration with architects Dave Pigram (University of Technology Sydney) and Niels Martin Larsen (Aarhus School of Architecture), who specialise in digitally based generative techniques and computation.[2] The objective was to test the casting technique's feasibility as a means of constructing a complex concrete grid-shell and further integrating computation in the overall form generation.

As inspiration for the overall form, we chose compression-only vaults in order to utilise concrete's high compressive strength. The concept of concrete vaults has been impressively exploited by Heinz Isler[3] and Felix Candela[4], who built numerous form-found, compressive vaults throughout their careers. These projects were all cast in situ, and needed to be defined by simple geometric rules, thus limiting their ability to adapt to different sites.

To enable the design of a compression-only structure constructed from mass-customised concrete elements, funicular form-finding techniques based on dynamic relaxation, previously developed by Dave Pigram and his colleague Iain Maxwell, governed the overall form of the structure.

We initiated the process of generating a compression vault from concrete elements by deciding to work with an open grid-shell structure, rather than a closed surface shell, in order to arrive at concrete components which are in essence a scaled-down version of the PETG cast beams and columns developed in the first design experiment.

### Geometry

When translated into a wireframe mesh, a double curved surface can be re-meshed from triangles into nonplanar hexagons, which reduce the complexity of each element. In order to utilise the circumstances in which the concrete can assume complex forms, we found it practical to design the components around the nodes and let two components meet on the line between nodes. The resulting concrete components are y-shaped with variable arm lengths and angles. A two-dimensional mesh was drawn and subjected to a dynamic relaxation algorithm, which generated a new mesh in pure compression. Volumetric, y-shaped representation of the concrete elements was parametrically defined on the basis of this mesh. These elements were then unrolled, laser cut, folded and assembled with rivets, reinforced and cast. The experimental pavilion was erected on top of a falsework made from recycled corrugated cardboard and then exhibited.

The established design method allows for a massive increase in the number of possible solutions that could be explored within a limited time span, when compared to physical models. Hundreds of simulations can be made in the design process, allowing the designer to develop a sophisticated understanding of the behaviour of the digital system and to create a structure which responds to various circumstances in the design brief.

### Notes on Buildability

The design experiment suffered from being complicated to erect. The need for an extensive falsework made assembly laborious and limited the saleability of the system. In other words, the overall form failed to be broken down intelligently into buildable parts, utilising well-proven construction methods in which the ring forces are utilised during construction. We are currently addressing this[5] by introducing a system of post-tensioning, which will stabilise the concrete during construction and help to cope with point loads in the finished structure.

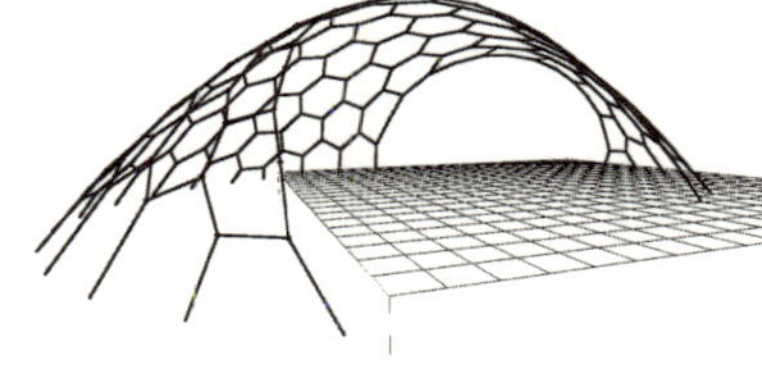

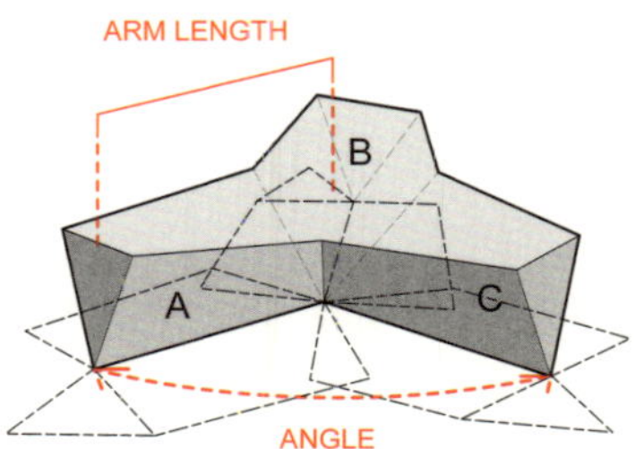


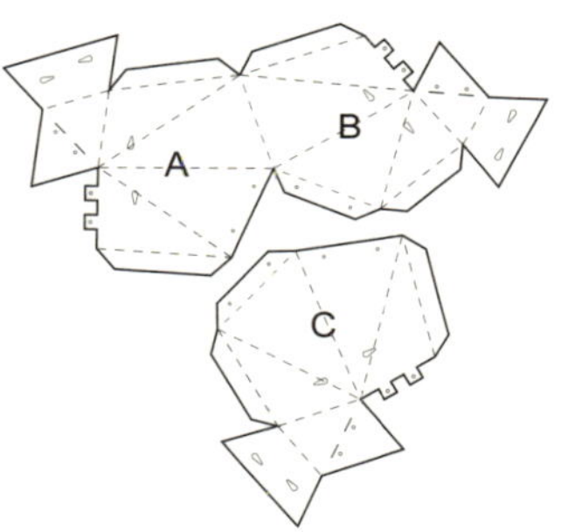


Definition of a double-curved surface by means of hexagons. © Ole Egholm Pedersen

Mesh after dynamic relaxation. © Niels Martin Larsen

Parametrically defined, y-shaped components and corresponding template. © Ole Egholm Pedersen

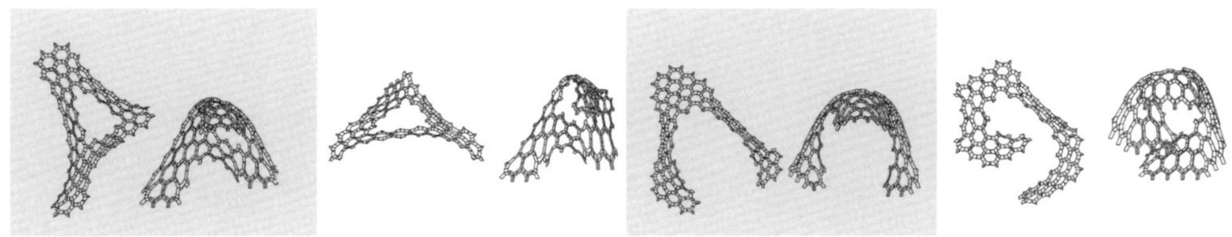

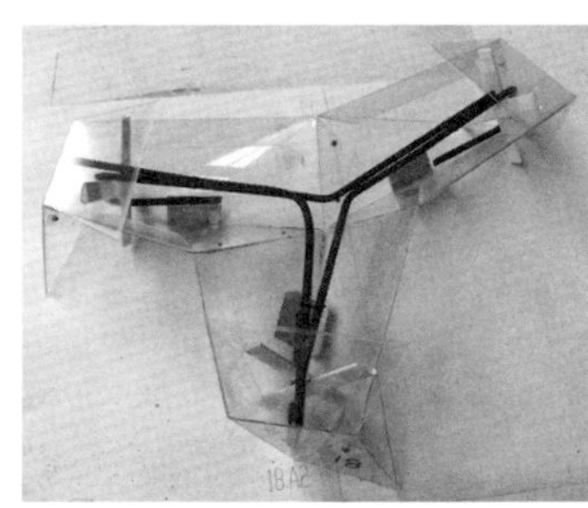

Top: Assembled casting mould.
Bottom: Mould filled with concrete.
© Ole Egholm Pedersen

Various design proposals.
© Ole Egholm Pedersen

Detail. © Niels Martin Larsen

Exhibition. © Niels Martin Larsen

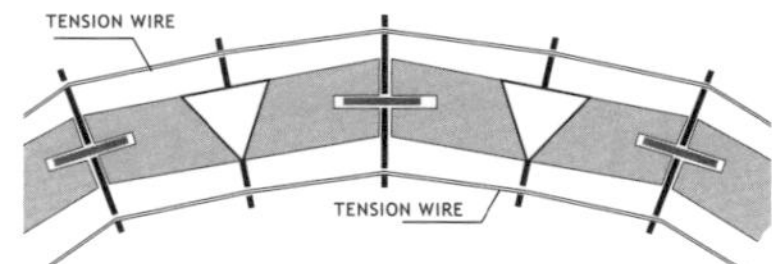


A principle for post-tensioning of the complex concrete grid-shell. © Ole Egholm Pedersen

**The Concept of Mould Tectonics**

With a tectonic approach to mould making the properties, characteristics or performance of material and technique influence one another and give rise to the form of the mould. This attention to the *relationship* between the *material*, *technique* and resultant *form* of the mould is pivotal in order to achieve moulds that are tectonic. Based on the findings in the design experiments and the above conclusions regarding material and technique, I present a series of dogmas which may provide the architect with tools to qualify concrete designs.

First, tectonic considerations should highlight resource optimisation: for instance, using a technique that does not require to be modified in order to allow for mass customisation when producing concrete casting moulds. Second, working with tectonic relationships between material and technique should inform design decisions and thus qualify the architectural performance of the concrete that is cast in the moulds. For instance, one might consider the possibility of manipulating the mould material into complex geometry, possibly utilising computation, and ways of achieving a closed, high-quality surface, first in the mould and subsequently in the concrete. Also, one should consider how the applied mould making material and technique can be read as imprints in the concrete. In the case of the concrete grid-shell, this reading is made possible because the rivet holes and crease marks are transferred to the concrete elements.

In conclusion, I present the following statement regarding mould tectonics:

*A mould with tectonic qualities may be conceived by means of investigating the potential of a specific* technique. *This can be achieved by the technique necessitating restraints of alterations to the mould, or by adding an extra layer of meaning.*

**The Concept of Concrete Tectonics**

The analysis regarding *mould tectonics* can be assessed in a larger context of *concrete tectonics*, leading to a series of statements proposing an attitude towards the act of making forms in concrete.

In this context, mould tectonics now constitutes a part of the *technique,* as stated in the introduction. As with the mould tectonics, each category (material, technique and form) may be assessed individually, in order to clarify their respective properties. It follows, with regard to the *technique*, that the analysis of mould tectonics presented above serves as such an assessment. Regarding the *material,* concrete, an account of all negotiable properties is outside the scope of this article, but two properties are highlighted: the material's inherent amorphous qualities; and its ability to cope with the hydrostatic pressure of concrete. As the form is a result of the material and technique, it is not assessed separately. Instead, the attention is once again on the *relationships* between material, technique and form. These relationships stem from the tectonic triangle presented in the introduction and include the mould tectonics, which is now part of the technique.

As with the mould tectonics, I propose a series of dogmas between the

Test of post-tensioning system. © Niels Martin Larsen

concrete material, casting and building technique, and the geometric form of the concrete structure. These are dogmas that allow the choice of technique to inform the overall idea about the forms in concrete, rather than formally conceiving a concrete geometry which is rooted in considerations that are purely spatial, and then leaving it up to the engineer or contractor to propose a casting technique.

In the case of the concrete material, it is worth considering how the technique could be presented in order to take into account, or even express, the hydrostatic pressure and amorphousness of concrete that affect and inform the architectural concepts of form and space. On a more pragmatic level, it is also worth considering how an overall geometric form could be broken down into buildable parts to accommodate the given technique, whether as elements or in-situ castings. Finally, maybe the concrete construction could reveal the mould tectonics: that is, ensuring that the concrete and its genesis constitute a visible and active part of the reading of the building or structure.

In conclusion I propose a statement that attempts to express *the tectonic complex of concrete*:

Concrete structures are tectonic when the concrete and mould materials, as well as techniques, have a significant impact on the initial idea about form, in such a way that the final structure can be said to be a consequence of material and technique.

**Ole Egholm Pedersen**
is an assistant professor at the Aarhus School of Architecture. In 2013 he defended his PhD thesis, titled *The Tectonic Potentials of Concrete*. In his PhD project he has been developing casting techniques with a close connection between the material, the technology and the form of the concrete element in order to reduce the use of resources and enrich the tectonic qualities of concrete constructions. The project has been financed by the research project "Towards a Tectonic Sustainable Building Practice".

Detail of cast prototype. © Ole Egholm Pedersen

1 __ Ole Egholm Pedersen, "Material Evidence in a Digital Context: Exploring the Tectonics Potentials of Concrete," in *Aarhus Documents 02: Context 2010/2011* (Aarhus: Arkitektskolens Forlag, 2012), pp. 14-15.

2 __ Niels Martin Larsen, Ole Egholm Pedersen and Dave Pigram, "A method for the realization of complex concrete grid-shell structures in pre-cast concrete," in *Synthetic Digital Ecologies*, ed. Mark Cabrinha, Jason Kelly Johnson and Kyle Steinfeld (San Francisco: Acadia, 2012), pp. 209-216.

3 __ John C. Chilton, *Heinz Isler: The Engineer's Contribution to Contemporary Architecture* (London: Thomas Telford / RIBA Publications, 2000).

4 __ David P. Billington and Maria Moreyra Garlock, *Felix Candela, Engineer, Builder, Structural Artist* (New Haven: Princeton University Art Museum, 2008).

5 __ Ongoing research by Ole Egholm Pedersen, Niels Martin Larsen, Aarhus School of Architecture.

Charlotte Bundgaard

# Readymades Revisited

## Industrialisation as a Contemporary Condition

In our industrialised world architecture is increasingly based upon industrially produced building products. Whereas industrialisation was previously related to production methods, which depended on standardisation and repetition, it is now rooted in advanced IT and high-tech manufacturing processes. We employ production machinery that is flexible and adaptable, and that no longer requires the repetition of completely identical elements. By virtue of IT-based production methods, building components can be unique and individualised, which means that the opportunity arises for creating a far more heterogeneous and expressive architecture, full of character, open to the wishes of users and to changing demands over the course of time. We hereby make a decisive leap away from what was traditionally considered to be the essence of industrialisation: namely standardisation, repetition and uniformity. The digital possibilities of creating a universe of industrial production, based on individualisation and one-off design, can offer a fundamentally different starting point from that of the previous wave of industrialisation.

Our understanding of serialism and reproducibility changes concurrently with the potential of production machinery to produce batches of objects which can be completely different, completely identical, or any combination of these, but, in principle, at the same price. This has a fundamental impact on the modernistic way of thinking about standardisation as a moral imperative, which implies that, based on a principle of equality, one can guarantee everyone the ownership of a house, a car, or access to light and fresh air. The possibility of individualising industrially produced elements creates new conditions; the economic, technological and ethical conceptions involved in serial production are undergoing a change. Individualisation can almost be considered a condition of our time. Both on the human and social levels, standardisation has been replaced by diversity and the need for individuality.

### Enriched Architectural Readymades

The idea of developing series of building components that are not identical, but on the contrary possess individual architectural character and material identity, challenges the traditional view of industrialised building elements as being monotonous, machine-like and architecturally neutral. Juhani Pallasmaa has criticised standard structures for their flatness and weakened sense of materiality. According to Pallasmaa, machine-made materials do not convey their material essence or age; on the contrary, buildings in our technological age aim at an ageless perfection. He also underlines, however, that recent developments point towards a new sensibility, which nurtures transparency, reflection and juxtaposition and thus creates a sense of spatial denseness, which may take over from an immaterial and flat technological architecture.[1]

The opportunity to create industrialised architecture that is not based merely on the rational repetition of identical components leaves the field open for the development of a far more heterogeneous and open approach to contemporary industrially based architecture. Building components can potentially take on different architectural roles within the building, and there is no longer a need to keep prefabricated sets of parts completely prefabricated. On the contrary, highly industrialised building components can be combined with totally different building systems, with in situ structures or with craft-based elements.

### Strategies of Juxtaposition

If we consider building components as being potentially still more distinct and individual, we face a design-methodological and architectural challenge. It might be fruitful to make a short digression into the field of arts, especially to the Cubist, Surrealist and Dadaist artists in the period around the 1920s, in order to discover how they dealt with the juxtaposition of elements. The concepts of montage, collage and assemblage in the field of modernist visual arts are not only manufacturing techniques, but also active artistic strategies. They all describe the design-methodological procedure of assembling units to form a new whole. This new whole consists of fragments taken from many different contexts, which are combined to form constellations and which deploy juxtaposition to create new meanings. Each fragment is part of something bigger, and carries meaning in itself: i.e., it is equipped with meanings by the context it has been taken from. But, by virtue of its meeting with other fragments, as a result of employing discontinuity and abruptness as design principles, the original meaning of the fragment dissolves and assumes a new role. In glimpses, the fragment suggests its real substance and contributes to the composite statement of the work. A closer look at collage- and montage-based works of art reveals certain design-methodological traits. Apart from the fragment itself, the way that the fragments are lined up, the way they meet, the space between them and the physical joints are all important points in the artistic process.

Still within the field of arts, the concepts of *objets trouvés* or "readymades" lead us even closer to the industrialised paradigm. The readymade, best known from the works of Marcel Duchamp, is an industrially produced object that is selected and displayed in a new context: the world of art. The artist has neither conceived, constructed nor fabricated the object. He has no control or contact with it during the fabrication process. He has simply selected the object among a never ending stream of industrially produced objects and, by selecting and displaying it in a new space, the art space, he consolidates its role as a work of art. So the artist does not conceive, design and manufacture every element himself. His most important task is the act of selecting and juxtaposing.[2]

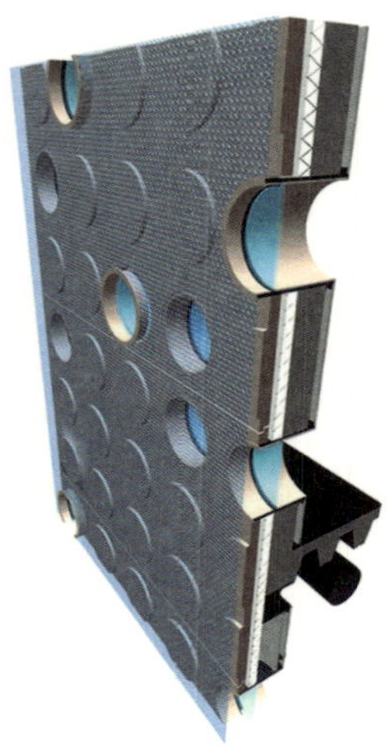

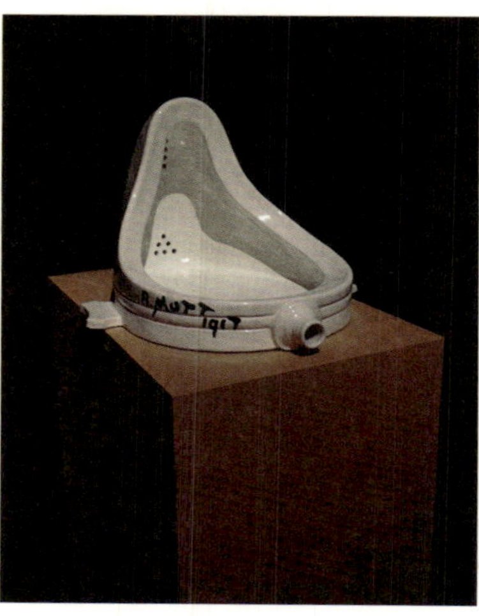

This façade component interprets the Venetian façade's texture and play with light. The component was developed by Ben van Berkel and Caroline Bos, UNStudio, in connection with a project for the College of Architecture in Venezia (1998). © UNStudio

Marcel Duchamp's *Fountain* (1917). Duchamp chose an anonymous object, signed it, gave it a title and exhibited it. A readymade was created. © Charlotte Bundgaard

When Lacaton & Vassal use greenhouse structures, they do not do it for the image, but for its use value relative to cost. It provides a compelling readymade space in an inexpensive house. © Philippe Ruault

### Lacaton & Vassal

The French architects Lacaton & Vassal introduce industrially produced building systems in their buildings. They line up prefabricated building parts, which they have not designed themselves, side by side with their own architectural solutions. They make architecture out of readymades.

One of the most famous examples of this practice is the Latapie House in Floirac (1993), where a prefabricated greenhouse structure forms a large part of the built volume. The house is situated in a housing-development context, and it was built on a very small budget. Nevertheless, it offers its inhabitants a perfectly integrated housing space for living their everyday life. The use of industrial construction principles allows for a building of great volume at a low price. In this case the architects discovered that greenhouse manufacturers have developed and optimised their products over a long period of time and thus present the best and cheapest solutions to the demands. The technical performance is high, and greenhouse manufacturers offer computer controlled elements in both the façade and the roof, ensuring natural ventilation and temperature control.

Lacaton & Vassal choose the greenhouse, not because of its narrative or pictorial strength, but simply because of its use value in relation to costs. They strive for the most effective and cheapest solutions possible. They fabricate a readymade space, which only demands limited elaboration, such as insulation, before it can be put to use. Lacaton & Vassal manage to create a compelling space open to other ways of living than what is found in common housing spaces. With limited means they make indoor spaces with diverse qualities: insulated or open, heated or unheated, transparent or opalescent. This variety provides different sensations and different usage throughout the day and the year. The act of living is crucial and, instead of concentrating on the aesthetic appearance, they consistently work from the inside out. They start by defining desirable ways of living and tailor the house like clothes in layers around the spatial activity. Which is not to say that they are not interested in aesthetics: Anne Lacaton talks about "Overt Beauty", the Beauty of the Obvious.[3] This reflects the architects' efforts to discover the most reasonable and clearest solutions, the simple being the most beautiful. They are not interested in the iconographic aura of the house, and generally they are sceptical about the contemporary cult status of the object. They are more interested in developing new spatial potentials. They do not focus specifically on the extra space, but more on the extra potential, life and experience which the extra space provides. They call it "surplus value".

### Technology as a Readymade

Lacaton & Vassal deal with technology as a readymade; they include technological solutions, developed in other contexts, in order to fulfil certain tasks. To them the architectural potential of technology does not lie in its original definition, but rather in its ability to be reprogrammed and combined with other things. Technology plays an important role as a constituent element in their architecture, but they do not worship technology for its own sake.

*"Haven't we lost our heads a bit about what it brings? Technology is interesting when it's a source of pleasure, usage, comfort and efficiency, but surely not as decorative performance."*[4]

They appropriate industrially produced products, readymades, to fulfil their architectural aims. They attempt to take as much advantage as possible of the character and performance of the readymade, altering or adding as little as possible. Through a deliberate displacement Lacaton & Vassal turn a greenhouse into a dwelling. This approach recalls exactly the artistic process of alienation within the modernist arts.

The habitable part of the house can vary according to the seasons. © Lacaton & Vassal Architectes

### Construction and Construed

Lacaton & Vassal manage to create compelling and narrative architecture by means of rather pragmatic industrialised building elements. To them the window is just a window; they do not add any other attributes beyond the result of assembling particular elements chosen from a large range of industrial products. Suddenly poetic dimensions arise from the pure addition of silent elements. In this way the architects manage to add a narrative layer to a pragmatic gesture.

This close link between the creation of concrete solutions and the creation of meaning forms the core of tectonic thinking. So tectonic thinking is not only about portraying a constructional logic. Tectonics is the creation of material realities, which reveal narrative meaning. Tectonics is about constructing with cultural references. This specific relationship between matter and meaning is clearly expressed by the Italian architectural theorist Marco Frascari. He claims that the concepts of "construction" and "construed" have to be united to give significance to architecture. Construction relates to the physical act of building,

The structural and material character of the greenhouse space merges with the expression of the adjacent living spaces. A simple and naked architectural approach is expressed. © Philippe Ruault

of assembling building elements, while construing is about creating meaning.[5] According to Frascari, both dimensions have to be present in meaningful architecture. Thereby he emphasizes that architecture cannot simply be construction without also containing a narrative layer; neither can it be pure narration without also having a physical, material manifestation. The actual construction of the building structure needs to contain a narrative layer in order to achieve a construing of the meaning embedded in the structure.

**The Tell-the-Tale Detail**

Marco Frascari points at the detail as being the meeting point between the mental construing and the actual construction. His notion of "the Tell-the-Tale Detail" describes the detail as a bearer of meaning.

*"This tension between construction and construing of an architectural object is solved in the architectural details, which are seen as the semiotic tools for the critical reading of architectural narrations."*[6]

"God is in the details". This expression is often ascribed to Mies van der Rohe and reveals the great importance of the detail in architecture. Detailing is the art of joining materials, building components and elements, in a functional and aesthetic manner. Detailing is where the architect expresses values and meaning in the architectural work, and this is where he can achieve innovation and invention.

The detail, or the joint, not only says something about the assembly of building parts, about how different materials meet, or about the production procedure, but also reveals the attitude, the way of thinking, behind the actual building. Frascari elaborates on the importance of the detail through references to, amongst others, Leon Battista Alberti, who describes his concept of *concinnitas* (adjustment) as an instrument for obtaining "beauty". Here the detail is an integral, constituent part in the creation of the coveted state of harmony. As a more contemporary example Frascari draws on Carlo Scarpa's "adoration of the joint" as the perfect realisation of Alberti's concinnitas. Every detail created by Scarpa tells the story of its creation, position and proportioning. The designing of the details is the result of a thorough study of the function of the detail. Scarpa's joints solve not only practical function, but also historical, social and individual function. When Frascari uses Alberti's concept of concinnity, he elevates the meaning of the detail from being merely a part or a unit, to becoming a hinge, revealing relations and meaning.

For Marco Frascari the detail is a fundamental architectural element, which tells the "story", the tell-the-tale detail. Frascari emphasises the architectural accordance between the detail and the building as a whole, using Alberti's quest for harmony or Scarpa's worship of the detail as ideals. But with the architecture of prefabrication this accordance is likely to change. Instead of manifesting coherence throughout every level of the work, the details of an assembled house might rather reflect a state of rupture. The prefabricated architectural work consists of a number of individual building components, produced in different contexts. Each of the components bears a certain approach to detailing. The architect might not create a homogeneous coherence, but might rather tell the story of the very act of juxtaposing deliberately different elements. Instead of expressing meaning through the detailing of each building element, the architect can express his agenda through the joints that link the components together. Roughly speaking, there are two approaches to the joint's expressing the act of juxtaposition: the exposed joint, which clearly expresses its function as a physical hinge, connecting one element to the other; or the absent joint, the dis-joint, which does not even try to create a link, but simply juxtaposes the elements without any visible links.

### Sustainable and Adaptable Readymades

Readymades carry significant potential in terms of the challenges we are facing right now. Sustainability, and the need to minimise the use of resources, optimise energy and think in terms of life cycle systems are important aspects, which readymades can assist. Using industrial production methods and dealing with a specific product or building system make it possible to develop still more accurate and advanced solutions. The development and production of building components or systems in a controlled industrial environment equipped with advanced digitally based tools offer the opportunity to optimise the use of materials and to refine the details.

This is exactly what Lacaton & Vassal exploit when they deploy greenhouse structures as specific solution spaces for their buildings. They choose the greenhouse product in question because it offers advanced climate control and ventilation systems, thus making it possible to live in the greenhouse spaces most of the year. Meanwhile, because the technical development and optimisation of the product take place within the industrial realm, the product is relatively cheap and ready to use. Moreover, the low price, combined with the fact that the producers keep on refining and optimising their products, makes it realistic and desirable actually to replace the whole readymade or parts of it after a period of time. The possible replacement of readymades is another characteristic trait, which addresses a topical issue: flexibility and adaptability. There is a strong need to support openness and the ability to adjust to new situations and challenges over time. Lacaton & Vassal tackled this in the Latapie House, where any possible way of living can be fulfilled in the raw, multi-greenhouse space.

Then what happens to the detailing of readymades in sustainable and adaptable architecture? The producers of greenhouse structures are continuously improving their products in order to meet increasing demands for sustainability, regulations, function, use and so on. This means that both the product itself and its details are likely to be subject to changes over time. The architect adopting readymades only rarely gets involved in the specific detailing within the actual readymade, as he or she selects the readymade object or system as a total solution. Nevertheless, the architect does adjust the readymade in order to fulfil specific tasks. Through collaboration with the producer the architect fits the system to the actual situation and context. Most likely it is a matter of adjusting lengths and dimensions, while the basic "grammar" of the readymade system remains stable. Marcel Duchamp's urinal and other readymades from the world of art are complete, "closed" objects, finite statements. The readymades in current industrially based architectural practice are no longer finished objects, but potentially objects in a state of continuous change, having to meet the ongoing challenges of sustainability and other important contemporary tasks.

Horizontal and vertical surfaces, dark and light tones, rough and smooth materials meet each other in articulated connections. Scarpa's details are never fortuitous or forgotten. They are always precisely positioned and proportioned. © Charlotte Bundgaard

But openness and adaptability demand a deliberate detailing, not only of the readymade itself, but also of the joints between the readymade and the remaining building systems. The details become the important link between the stable parts and the flexible elements. However, in the case of Lacaton & Vassal, the architects do not clearly distinguish between readymade elements and other building systems in use. In the Latapie House one cannot clearly tell the readymade greenhouse structure from the rest of the building elements. The structural and material character of the greenhouse space merges with the other living spaces: probably because Lacaton & Vassal want to express a simple and naked architectural approach.

Is this because the readymade way of thinking is now an even more integrated approach to industrially based building? Because of our options for working with IT-based production methods, is the readymade no longer a fixed and stable object like Duchamp's piece of art, but rather a shapeable element in motion? Or is it simply because Lacaton & Vassal celebrate a modest and homogeneous aesthetic, where there is no need to expose differences between readymades and non-readymades? Is it that the readymades form the basic condition of the building and, when they are distinctly important, fill and colour the whole building?

### Detailing an Architecture of Readymades

As far as Lacaton & Vassal are concerned, God is obviously not in the details. Their architectural approach, celebrating readymades, cheap materials enhancing simple aesthetics, and openness to change, contradicts the idea of developing details with an aura of perfection. When deliberately deploying prefabricated building systems, one simultaneously gives up the idea of controlling the details on all levels. Lacaton & Vassal underline the importance of using prefabricated elements in order to achieve low costs. This would be an impossible task if they insisted on reaching coherence in the detailing by altering and adding their own details to the prefabricated systems. And they probably do not even strive for aesthetic egality in their works. To them, readymades bear their own rational aesthetics, which might very well be different from those of other parts of the building. Nevertheless, the Latapie House seems to have a coherent architectural expression, a certain homogeneity, which is based upon naked solutions and simple materials. In the entire house the architects match the tone of detailing, which belongs to the prefabricated greenhouse structure. Instead of adding a new layer of detailing to the industrial building products, they share the rational and industrial approach, distributing it throughout the whole building. Hereby they achieve a simple aesthetic, which tells a story, but not the story of harmony and perfection. Their story is more about raw and cheap solutions creating rich and stimulating spaces.

However, Lacaton & Vassal do not simply leave the readymades "as found". They orchestrate their buildings with elements of everyday architecture. At first sight most of their architecture resembles sheds or barns: until one notices the precisely dimensioned openings, the studied combinations of structural elements, the accurate use of materials in the whole and the parts, and, last but not least, their quiet, modest, yet convincing detailing.

Thanks to Marco Frascari's notion of the tell-the-tale detail we understand that detailing is not just a passive act of solution, but also an active tool for the architect, both a method for establishing reflection and a tool for realising physical form.

*"Detailing is not an abstract passive combining and joining of details, a composition, but rather a method of productive reasoning, a project of possible material realities."*[7]

In a world of prefabrication, where architects juxtapose building elements from different contexts, the details are also an important means: on the one hand to express a distinct architectural approach; on the other, by means of the joint, to literally assemble the elements. Lacaton & Vassal do not accentuate the juxtaposition of the greenhouse structure and the other parts of their buildings. They do not search for dramatic collision or clear distinctions between the readymades and their own architectural web. Instead they seem to tell the story of a relaxed approach, with room for both construction and meaning.

*"Together, they suggest that architecture is the intersection of a site, a brief and restrictions that must be efficiently, functionally and rationally satisfied, but there must also always be something more that eludes efficiency, that is pleasure, the unexpected, the poetic and that, in the end, is essential."*[8]

**Charlotte Bundgaard**
is an associate professor at Aarhus School of Architecture. In 2006 she defended her PhD thesis, focusing on montage as a tectonic strategy within industrialised building construction. Since then this focus has been the focal point for her research and teaching. Charlotte Bundgaard is the author of a large number of papers and articles. In 2013 she published the book *Montage Revisited – Rethinking Industrialised Architecture*, based on both her PhD thesis and her recent research.

© Lacaton & Vassal Architectes

1 __ Juhani Pallasmaa, *The Eyes of the Skin. Architecture and the Senses* (Chichester: John Wiley & Sons Ltd, 2005), pp. 31-32.

2 __ An architectural strategy for industrialised architecture based on the concept of montage is developed in the book: Charlotte Bundgaard, *Montage Revisited – rethinking industrialised architecture* (Aarhus: Arkitektskolens Forlag, 2013).

3 __ Anne Lacaton & Jean-Philippe Vassal, *Il fera beau demain* (Paris: L'Institut français d'architecture, 1995).

4 __ Anne Lacaton & Jean-Philippe Vassal, "A Conversation with Patrice Goulet," *2G, Lacaton & Vassal* 21 (2002), pp.122-143.

5 __ Marco Frascari, "The Tell-the-Tale Detail," *VIA 7, The Building of Architecture, Architectural Journal of the Graduate School of Fine Arts* (Pennsylvania: University of Pennsylvania, 1984), pp. 22-37.

6 __ Marco Frascari, "The *Particolareggiamento* in the Narration of Architecture," *JAE* 43/1 (1989), pp. 3-12.

7 __ Frascari, "The *Particolareggiamento*," pp. 3-12.

8 __ Anne Lacaton & Jean-Philippe Vassal, *Lacaton & Vassal* (Paris: Éditions Hyx/ Cité de l'architecture & du patrimoine, 2009), p. 20.

# 3

## Construction Materialised

*The deep understan-
ding of the potentials
of materials informed
by the knowledge
embedded in traditional
building practices can
form the basis for new
innovative solutions*

# Introduction

*All materials have their own inherent potentials and qualities. How can we as architects ensure that an optimum release of these potentials becomes the core element of the architectural creation?*

*A deep understanding of materials is often embedded in local building practices. How can this knowledge and tradition be maintained and developed further in a more and more globalised and industrialised way of organising the building processes?*

One of the core elements of tectonic thinking and practice is a deep understanding of materials and the way they are processed and incorporated into the final design of the building. The study of materials and their properties is within a tectonic perspective a fundamental part of the education and work of an architect. This way of practicing is just as relevant in the use of new as in the refinement of traditional materials. When the potentials of the materials used in the construction are released, the tectonic appearance of the building not only tells the story of the building's own making; it also offers an insight into the logic of the materials used. When the potentials of the construction are exposed they can form the platform for the future use and the further development of the building. In this way the very logic of the construction is materialised and becomes the core of the architectural expression of the building.

When studying regional building practices, we find that a profound insight into the properties and use of local materials is passed on from generation to generation. This knowledge forms the core of the crafts and the building culture of a region – often not as explicit knowledge but more as the craftsman's inherent skill – embedded in the practices and building details used. The contemporary construction industry is getting more and more industrialised, digitised and globalised. In this process of rethinking the way we construct both the processes and the building itself we might be neglecting the valuable insight and meaning of our regional building cultures. A huge potential lies in the study of our traditional building methods, transferring the knowledge of these fields into modern ways of constructing. This could qualify and enrich the development of innovative solutions solving some of the challenges within contemporary architecture and construction.

**Positions and Perspectives**

The four texts in this chapter introduce different perspectives of the close link between material, technology and construction. The first text, by Bijoy Jain, is a profound introduction to the unique work methods of Studio Mumbai – methods based on a deep understanding of the connection between materials and craftsmanship. The close link between material and man is also the point of departure of Thomas Bo Jensen's manifesto arguing for the poetry of brickwork. The third text, by Johan Celsing, likewise focuses on brickwork and how different characters and qualities of the material have been utilised in projects by his office. Finally, Jonathan Hale discusses how we as humans engage with technology and even absorb it into our body image.

The four positions can be recapitulated in the following short descriptions of the different perspectives of the contributions to the discussion:

The Indian architect Bijoy Jain describes the practice of the architect – whether tangible, ambiguous or theoretical – as primarily concerned with the nature of being. This ontological understanding of the word "praxis" is both the title of his text and expresses the work of his studio and workshop, Studio Mumbai. In the studio's work, architecture is created from an iterative process, where ideas are explored through the production of large-scale mock-ups, material studies, sketches and drawings to form an intrinsic part of our thoughts and body. The studio's projects make a conscious commitment to the environment and culture by the physical and emotional engagement of the people involved in the building process. The specific character of the work methods of the studio is described in a detailed exposition of recent projects, with a special focus on the Copper House II project.

In the text "The Poetry of Brickwork" the Danish researcher and architect Thomas Bo Jensen addresses some evident barriers in contemporary architecture, in making tectonically pronounced and clear brick architecture. The text points out some of the qualities that have made brick a culturally sustainable building material: its specific physical characteristics linked with the hand and the body, and the symbolic spiritual significance of brick, rooted in the fact that countless generations have lived with it. The text can be seen as a kind of manifesto stating the need for renewing the bond between man and material if we are to work again in accord with the sustainable qualities of clay and brick.

"Gravity and Clemency" is the title of Swedish architect Johan Celsing's detailed description of three recent projects designed by his office. The architectural intentions of the projects are seen in a tectonic perspective. The buildings explore different ways of working with brickwork – from the gravity of the masonry seen in the Church of Årsta to the lighter use of brickwork in the New Crematorium of the Woodland Cemetery. The text discusses how the use of materials is to be seen in close connection to the topography and cultural context characteristic of every individual project. The specific detailing in the work of the office is often developed through working with full-scale models to understand the unique character of the selected solution. In this way the building process becomes an ongoing examination of the tectonic potentials of a given material.

In "Cognitive Tectonics: From the Prehuman to the Posthuman", the British architectural theoretician Jonathan Hale explores the relationship between the human and the technological by discussing a number of examples of the way in which we engage with technologies and incorporate them – literally absorbing them into our body image. This process features a number of important cognitive consequences, discussed in the text with references to theories developed by Maurice Merleau-Ponty, André Leroi-Gourhan, Raymond Tallis and others. This more theoretical discussion leads to the argumentation for a continuing relevance of tectonic articulation in architecture, for example, in the creation of engaging and richly layered environments that contain visible traces of both construction and occupation – spaces that invite engagement with both the bodies and minds of future building users. The New Art Gallery Walsall by Caruso St John Architects is used as an example of this way of thinking and practicing.

Ulrik Stylsvig Madsen

Bijoy Jain

# Praxis

*Praxis is the process by which a theory, lesson or skill is enacted, practised, embodied or realised. It may also refer to the act of engaging, applying, exercising, realising or practising ideas.*

All-encompassing, the practice of the architect, whether tangible, ambiguous or theoretical, is primarily concerned with the nature of being. This ontological understanding of 'praxis' may begin to express how the work at Studio Mumbai is created from an iterative process, in which ideas are explored through the production of large-scale mock-ups, material studies, sketches and drawings to form an intrinsic part of our thought and body.

At Studio Mumbai we endeavour to look at thought, idea and action as a seamless entity: eventually they overlap and become one. A question then arises as to how to manifest a thought or an idea as a physical expression. Projects are developed through careful consideration of place and practice that engages intently in the environment and culture, the physical and emotional engagement of the people involved, where building techniques and materials draw from an ingenuity arising from limited resources.

What is the motivation of architecture? Where does the motivation lie in the creation of architecture? These are the fundamental questions we continuously ask ourselves. To put things in place and to arouse curiosity has become fundamental for us. It has become a tool that allows us to build architecture. In our studio we draw and create at the same time. It is like a human, rapid prototyping system which, after the initial drawing, leads to the creation of an object half a day later – an object with which you can engage further. This includes full-scale mock-ups and scale models, tools developed specifically for the current purpose, instruments of representation that express the ethics of space. It allows for a community to engage; it is an extension of our collective internal and external world. The model provides an opportunity to perceive this.

We are currently building a house in bricks made out of compressed mud obtained from the excavation of the foundation on-site. It is an attempt to keep

Bird Tree, Rajasthan © Ruedi 2012

Top: At the workshop in Nagaon
Bottom: Postures in work © SMA 2007 – 2012

the resources we have collected and to reduce the movement of man and machinery in order to enable a focus on the immediate environment. The site is in Ahmedabad, where the climate is quite dry and where there is a tradition, both ancient and modern, of brick masonry construction that enables us to work in this particular way.

In order to test the methods and materials, we made a manual press in the studio, which precisely simulated, on a smaller scale, the bricks we would make on-site. The action of the model simulated the building method through which many of the full-scale details for the project emerged. A shift occurs in scale of the treatment of sand that can be transferred to a model. Both of the houses, one in the studio and the other on-site, were literally built simultaneously, one slightly trailing behind. The idea for this method is to overcome scale with the intention to retain the core value or ethic of concept, structure and space.

The model then becomes an autonomous entity with a life of its own. It is put out in the rain and the sun. We can then observe how the materials will weather over time. It is more than just an exercise in drawing or building. It is about identifying a sense of equal realm, a sense of position on a particular scale.

### Postures in Architecture

Can one's proximity to the work during the process of making influence what is being made? Taking a position as a craftsperson is no different from that of a painter or fencer. As in judo or other martial arts, it is all about composure – like a dance that creates a certain atmosphere. The embodiment of that atmosphere is the potential material for what is being made.

It is a complete system in which one engages in something intimate. This intimacy of the body that envelops what is being made counts for efficiency and an immediacy that engages the senses. What we have observed is that if, as a culture, we lose the ability to strike or maintain certain postures, we may lose the ability to produce certain kinds of work. The relationship between the mind, the hand and the body is altered.

Building the well © SMA 2004

### Tara House

This is a site on which we found an excavation in the ground filled with fresh water. In this region there is no formal infrastructure or system for the supply of water. One has to rely on what is underneath the earth. We decided to use this as the basis of the project, a sort of liquid foundation. It serves the house and its inhabitants. Water and its sustenance become the core value. This was what we wanted to put our energy into, this opening in the ground. The house then forms an enclosure around this belly that protects the water source.

The work demanded an understanding of the depth at which one will find water, the size of the fissure in the ground, and careful attention as to how to ensure a continuous flow of water within the space. With this consideration, we had to be mindful of the weight applied over the fissure, as over time this force might compress the soil and eventually inhibit the water flow.

There is a ritual to follow when one builds a house, in which one tries to find the water source. The well is the first thing one builds. This is done in the dry season, March to June, when the water level is at its lowest.

What is formed is a room for water. In the monsoon, the water goes up all the way to the ceiling, and it rains through the openings which were created to bring in air and light. It is a room that undergoes transformation. It registers time, seasons, the ebb and flow of the tide and the waxing and waning of the moon.

Do these ideas about tectonics relate to something deeper within our own internal space? Is it possibly a manifestation of our internal world expressed externally? Maybe it is neither one nor the other, but an oscillation between the two. What is essential is that one experiences the presence of water, the

Tara House in the summertime where you see the registration of the waterlines, which was a couple of hours later where it registered the high tide © SMA 2004

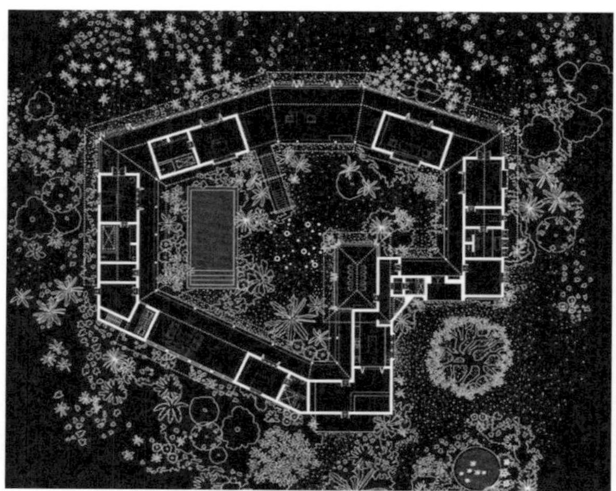

Tara House plan. © SMA 2004

Section through the well at Tara House. © SMA 2004

sea, with an inner intensity, greater than what one perceives with the eyes. This experience demonstrates how the denial of one of our senses heightens the others. We primarily use our eyes to observe the world. It is an interesting phenomenon, in which, as a result of this denial, we begin observing from a different point of view. It becomes an architecture that engages us. This potential was not anticipated, but at some point we became more conscious of mass, density, weight, light and air, which are registered through all the other senses.

**561/63 Saat Rasta**

How do we work with an intense surrounding while at the same time also considering the adjacencies? This is a project in the heart of the city, a run-down warehouse in Mumbai, bordering the railway tracks. The space is divided into a studio, workspaces, a library and houses.

Outside, along the railway tracks, there are some structures that have been here for forty to sixty years. They are no longer illegal, but are not officially on the map of Mumbai. They both exist and do not exist; they are known about, but wilfully ignored. In trying to preserve the quality of the space, what was of primary importance was how to negotiate these adjacencies. The original project had a gabled roof. However, to retain this we had to consider the rain and displacement of water from the large surface of the roof. Tolerance between the structure and the adjacencies was too tight to allow for gutters or water channels to enable water to be diverted efficiently without any obstruction. At some point we decided to slope the roof inwards, drawing all the water back into one space. Accordingly, the courtyard was formed.

We looked at ways of creating an even condition. We explored the potential of sharing the same phenomena equally, irrespective of the cost or unequal division, creating a shared condition of the volumes of air, light and water that enter these spaces. It works like a flute.

While walking along through the central corridor or street, as one's head and body gradually move through this space, looking through and into the courtyards, one senses different volumes of light and air and consequently differences of sound.

Our interest is not so much in the plan or in functionality, because at some point, consequently, the project can turn into a bookshop, a school or a hospital, etc. The programme is not important. It is the fundamental quality of space, light, air and sound or lack of it, entering through the courtyard, which is intrinsic to the space.

Our interest lies in the experience of phenomena. The construction of our projects is not so much about getting the work done precisely, down to the last millimetre, but more importantly about collecting and retaining atmosphere and, in doing so, life itself.

In this case we made a large-scale model with the sheet metal roof and wooden frame, which we then left out in the rain. While these models were

Aerial view of the site. © SMA 2012

Saat Rasta sketch and model exploration in brass. © SMA 2012

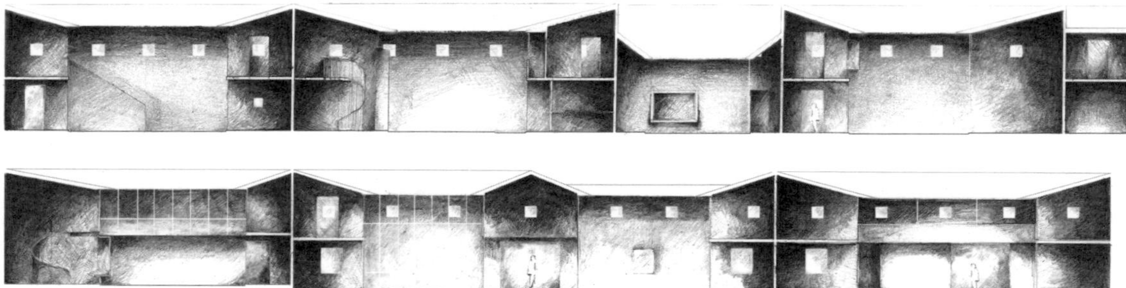

This drawing shows the quality of light and water and rain and everything that can possibly enter this space.
© SMA 2012

Aerial view of the model on-site.
© SMA 2012

being made, something did not seem right about the appearance of the models. What we found missing was the glazing, which formed the most important part of the project. We were looking for parts of fluted glass with very thin lines to use in the model. We tried to simulate this, but were unable to achieve the same quality. It was unnerving to see that what we were going to build was not what we had imagined. As in the Ahmedabad project, we needed to transfer the phenomena of the experience to the scale of the model. The studio came up with the idea of screen-printing on glass, and acid etching the glass surface. Gradually, through this process, we developed a model that simulated the experience.

The studio functions like a laboratory. The idea of models and drawings is to test the phenomena continuously, again and again, until they become part of our shared culture. Eventually, as the project is being built, there is no real need for drawings. It is internalised into our system. This method allows for tolerance or discrepancy, a natural built-in error that enters the work. As a result there is potential for the space to resonate.

Copper House II well.
© SMA 2011

**Copper House II**

The site work for this project focused on the conditions: Once again we started with the well that then provided the soil for the construction, and eventually water for the house. It was located by a water-diviner, who used copper sticks that moved in different directions as he walked across the site, until a mark was registered in which the sticks remained stationary and pointed specifically to a spot on the ground. The process is called "water divining". This is very specific work, because it involves an intuitive understanding of the environment and the ground below.

The soil was manually transferred with wheelbarrows and used to construct the high ground about a hundred meters away. It was done a month and a half before the rains in order to accurately record the lowest water level during the dry season. In the monsoon, the soil was compacted naturally by the rain, thus reducing the need for machinery on-site.

The inside and outside of the house are seamless in experience, not in the physical realm but, more importantly, in the experience of the atmospheric air that surrounds and inhabits the exterior and interior of the house. The house was located specifically to take advantage of the foliage of the trees, which varies from summer to monsoon. In summer the trees let in light and air, while in the monsoon season they act as a protective raincoat, as the density of the leaves increases due to the wet climate.

The landscape, especially in the monsoon, can be quite intimidating and claustrophobic. Living in such conditions, one begins to understand why certain architectural features have developed in this region. The plinth, which raises the house slightly above the ground, is a simple device that creates a space between what naturally inhabits the ground and how we as humans inhabit the ground. The courtyard works as a device to relieve the site, the house and the inhabitants of all their anxieties. Our way of building offers a great opportunity to use this potential. It is about-involving the collective resources of an environment engaged in a construction based on ontology and the optimum measure of place. The development of an idea through mock-ups, material studies, models and drawings works as an effective and economical process to reveal the environment we are engaging to construct. Each participant picks it up and develops it like a dance; back and forth in space, exploring the place. It does not matter how exact the dimensions are; it is more a question of how the spaces fit inside each other. The builders who are actually constructing the project make drawings, thus developing an innate understanding of the work to be done. This manner demands attention to oneself and to the act of making, as

Mock-ups and models at the workshop. © SMA 2010

Copper House entrance view. © Helene Binet 2011

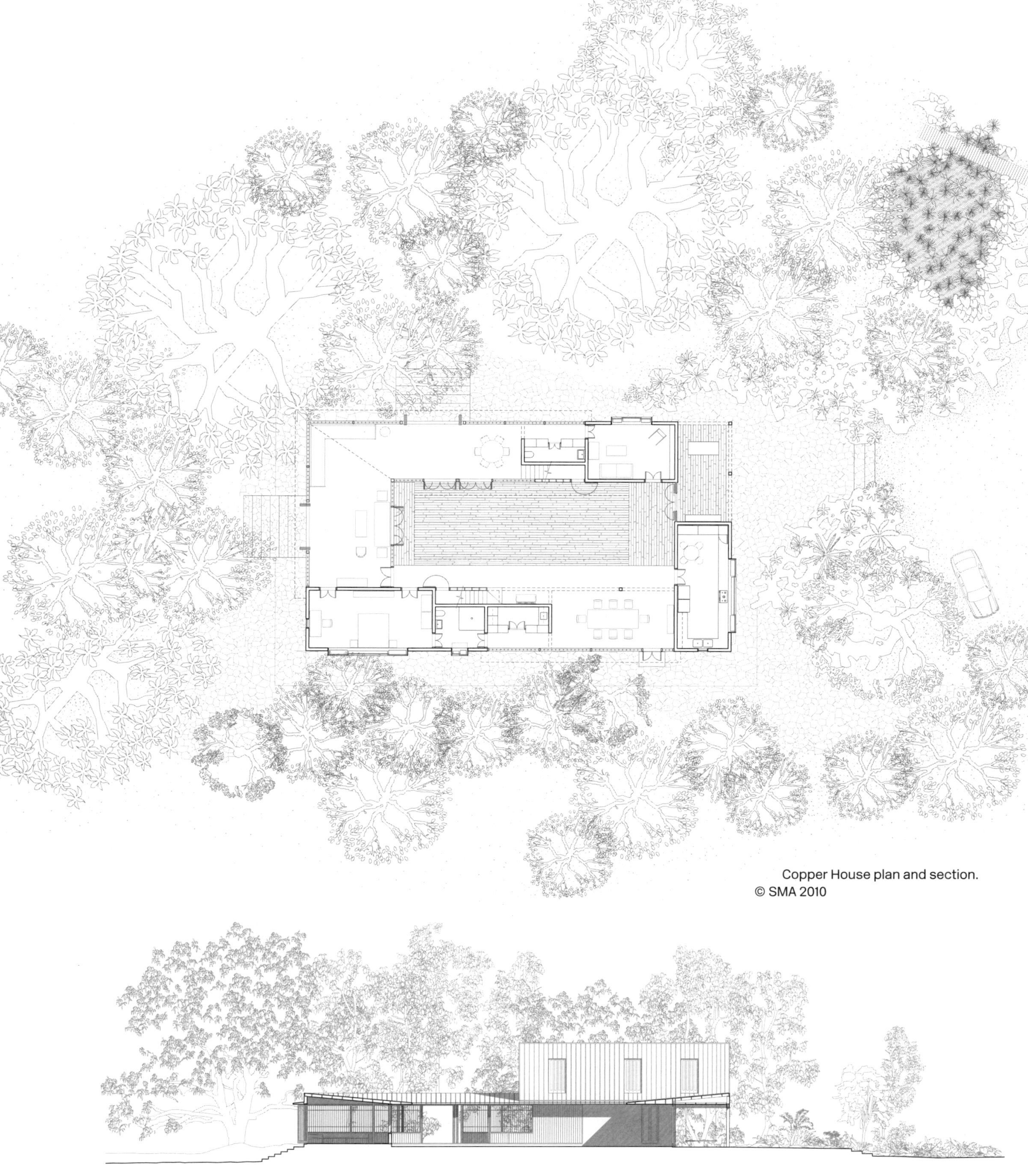

Copper House plan and section.

Left top: Copper House, courtyard © Helene Binet 2011
Left bottom: Copper House courtyard during the rain © SMA 2010
Right: Copper House, interior © Helene Binet 2011

opposed to a specific and predetermined set of instructions to be carried out.

The carpenters tend to engage with the work in a particular way. They are also involved all the way. They make a series of drawings while on-site or at the studio, working to produce things and refining the idea at the same time. Much of the communication is done through shared narratives. There is something comforting in the words, "Yes, we have a sense of where we are going" and then one starts developing stories. We build on these stories, regardless of location and origin. When we develop a project, our aim is to create something both familiar and unfamiliar. The idea is to create an environment where there is a connection for the people building it.

There is something about the changing of seasons that one anticipates. The passing of time and materials becomes very important. When architecture embodies them, it becomes an extension of ourselves. In a house built for rain, the enjoyment of experiencing the rain is part of the essence of the house.

The roof and first floor are clad in copper, which usually takes about twenty to twenty-five years to achieve a patina. In this case it will probably achieve a patina in five years, because of the climate. Looking at copper over a period of time relates to the idea of camouflage, a changing environment, in which subtle shifts occur, similar to those of our skin and the way in which things change within us.

When we finished construction, I felt that the house lacked something, something within. The rock entered the picture at the very end. The client said that her mother was doing some excavation in the south of India in the mountains and that she had some large rocks. A month and a half later the rock arrived, and we had no idea how we were going to get it into the house. We could get a big crane to transfer it over the roof, but it would be too risky: a chain might slip and break. We couldn't take that chance.

Finally it was decided simply to remove the doors along the joint and lift them out. About fifty people pulled a two-ton trolley through the entrance with the five-ton stone on it. The rock was eventually wheeled through the door, where the crane then picked it up, gently swung it around and dropped it in place. Thus the project was completed.

The placement of the rock in the courtyard then led to an interesting situation . While the idea of the courtyard was to relieve the anxiety of the ground and the pressure of water during the floods, the rock was now pressing down on the flood, where it would be likely to sink in water. In a sense it absorbs this intense landscape. Everything is concentrated in the rock and it somehow seems to have settled the project.

*The house will fade away and disappear in time. What will remain are these phenomena, which are integral for continuity and for something new. This could potentially be picked up where all of this was left off.*

**Bijoy Join**
is the founder and partner of the architectural practice Studio Mumbai in India. He read architecture at Washington University, St. Louis, and at SCI-Arch, Los Angeles. From 1989 to 1995 he worked in Los Angeles and London, both for other architects and independently, before returning to Mumbai, where in 2005 he founded Studio Mumbai. The practice has won international acclaim for its work, which is based on close collaboration between architects and craftsmen. The studio presented its special approach to the creation of architecture at the 12th International Architecture Biennale in Venice in 2010 with the project *Work Place*.

Section through Copper House and surroundings. © SMA 2010

Thomas Bo Jensen

# The Poetry of Brickwork

Modern brickwork is in crisis. Over the past few decades, we have increasingly failed to utilise most of brick's fine characteristics. A brick is still a brick, but the qualities that are given priority today have nothing to do with its density. Brick has become an irrational building component whose continued use depends on habit rather than on building physics. It might therefore be just a matter of time before brick is phased out in favor of lighter cladding materials. Is this simply a natural evolutionary development that it would be pointless to resist, or is it an expression of a general cultural derailment that characterises late modern societies?

The world is ailing. We have reached "the age of repair". Architecture has been proclaimed the biggest energy culprit, and no wonder, since buildings' energy accounts comprise a slew of subitems, from the manufacture and transport of building components to daily operations and household consumption. The treatment prescribed is a radical reduction in a building's permitted energy use. This strategy, born from a long tradition of treating only symptoms, can be compared to dosing animals at the top of the food chain with medicine that will alleviate side effects in the hope of curing the man-made dysfunctions of the entire ecosystem.

This essay points out some of the qualities that have made brick a culturally sustainable building material: its specific physical characteristics linked with the hand and the body, and the symbolic significances that are linked with the spirit and rooted in countless generations' coexistence with brick – a kind of symbiosis between man and material that does not exist with other building materials. It is time now to reaffirm this bond if we are to work again in accord with the sustainable qualities of clay and brick.

### Contemporary Brickwork

§1 __ For millennia, walls built of bricks have undergone a slow evolution in which the methods of constructing them have remained fundamentally the same. They are stacked structures, the bricks having been joined using mortar,

When made carefully, brickwork is like a piece of woven textile. Old wall in Roermond, Holland.

with the other building elements resting on the load-bearing walls. Brickwork has consequently often played the primary structural role. When buildings have reached the ruin stage, all that was left was the eroded remains of the brickwork.

§2 __ With the advent of cast iron and reinforced concrete in construction at the end of the 19th century, concrete and steel gradually encroached on brickwork's domains. This development accelerated beginning with the oil crisis in 1973, and many European countries introduced legislation that complicated the use of load-bearing brickwork. Today, load-bearing brickwork is a rare exception. Bricks are used primarily to cover other, more rational, building materials. In keeping with more stringent limits on energy consumption in recent years, facing walls have been "pushed farther out into the cold" to make room for more insulation.

§3 __ This means that many of brick's inherent qualities are no longer utilised. In addition to its load-bearing properties, they include thermic characteristics: the ability to "breathe" and stabilise humidity, to absorb and release heat in the course of the day, and to even out big fluctuations in temperature. These are factors that in a natural way help create a healthy indoor climate and a durable, almost maintenance-free structure. Thin facing walls have posed a number of technical problems. To solve them, strong cement mortars have replaced more flexible lime mortars. Strong mortars and a stratified structure, consisting of many materials that behave differently, have moreover necessitated the use of vertical dilatation joints filled with rubber that cut through the facing wall at suitable intervals. Modern brickwork has consequently been transformed into a composition of stiff sheets that are treated as if they were concrete elements.

§4 __ A world of images. The development described here has run parallel with an increasingly more image-driven world. The two-dimensional effects of images have infiltrated architecture, including contemporary brickwork, which because of its use of facing walls has been obliged to cultivate the aesthetics of the surface. Lars Morell wrote of the metamorphosis that occurs in architecture when the beholder sees it transformed from a physical mass into something metaphysical. "This transformation comes about differently in architecture and in painting", he noted. "In a painting it is in the colour, the composition, the brushwork, that the motif is transformed into something beautiful. In architecture it is light, and the weather in general, that strikes the actual, functional buildings and touches them with beauty. All art contains a transformation, but in architecture it is the naked, impassive shapes of the external world that appear transformed by the weather."[1] By forcing the external shapes of architecture to use the means employed in painting and photography, the uniform effect of the surface can be heightened. But in doing so, the

Top: The open-air-museum in Arnhem, Holland (1998): 140 meters of brick patchwork. As an answer to modern building technique, the dilatation joints play a natural role.
© Ole Meyer

Bottom: Contemporary brick cladding. Bricks are clearly separated from the primary construction.
© Thomas Bo Jensen

architect renounces the tectonic vocabulary's knots, cracks and shifts, which – with the aid of weather – can evoke architecture's metaphysical moments.

**Brickwork is Wickerwork**

§5 __ Dressing. In our age's "facing-wall reality", where buildings consist of many layers and where the bearing structure is often hidden, Gottfried Semper's *Bekleidungstheorie* (theory of dressing) has become current once more. It was Semper who first pointed out brickwork's similarities to wickerwork and woven textiles. Semper believed that when wickerwork walls were abandoned in prehistoric times for more durable walls of masonry, something remained: a kind of "skill" that persisted in the technical metamorphoses of the evolution of cultural history. The woven character of joints and bricks, imprinted in the surface of the brickwork like the pattern of threads in a garment, was visible proof to Semper of the continuity and cohesion of cultural history.[2] When a brick is put together with other bricks, it relinquishes its own singularity and becomes a mass and a surface, just as a thread becomes cloth. Semper viewed the way in which this happened not as an isolated new creation, but as the result of a technical skill that had been handed down through countless generations.

§6 __ The urge for order. Stacking and joining bricks into bonds can also be viewed as an expression of the urge for order that has followed human beings since the ritual and rhythmic ring dance sounded the key notes of civilisation. For Semper, this connection was an obvious part of mankind's existence in time and space.[3] When we place brick upon brick, we do so not in random order, but in carefully prescribed combinations that are intended to ensure stability and systematisation. This systematisation – this urge for order – produces patterns which, although they probably do originate in structural syntaxes determined by function, simultaneously create their own ornamental character independent of material and time. In Mari Hvattum's interpretation of Semper's *Bekleidungstheorie*, the wall is thus primarily "an ornamental memory of the loom's rhythmic process of order, and only secondarily a structural, bearing element". She further noted that what was important for Semper was "not the construction, but the ornamental envelope: the wall's symbolic *Bekleidung* as it had been transformed over the millennia from primitive wickerwork into a surface covered with mosaics or paneling".[4]

§7 __ The knot. Semper considered the knot to be the archetype of technology. In it, he saw the earliest technical expression of the human desire to create new significance by joining things. The joining of cords creates a link between two elements that are insignificant in themselves. The knot's twisting is moreover this link's necessary technical form. At the moment when human beings were able to tie a knot, they also became aware of the beauty of its soft bow shapes, Semper maintained. In his view, technology and aesthetics have always gone hand in hand. By considering the rhythmic structures in textiles and wickerwork from prehistoric times, Semper found a confirmation that technology never stood alone. There was always a relationship in which technology and aesthetics exerted a mutual influence – or in which technology challenged artistic awareness and vice versa.[5]

§8 __ Poetry. Semper's thesis was later supported by Martin Heidegger, who in his etymological investigation of the Greek origins of the concept of technology, *téchne*, noted the dual significance of the concept as both handicraft and art, and its close relationship with the *art of creation* and (through it) with the concept of *poiêsis* (poetry, bringing-forth).[6] *Téchne* is not only the designation for crafts and skill, "but also for the arts of the mind and the fine arts", noted Heidegger.[7] For this reason he warned against viewing technology as purely instrumental. If we do, we ignore what Heidegger understood as the "essence" of technology. For him, the essence of technology was not technological at all, but rather "the underlying understanding of being that makes technology possible".[8] This is why "the essential reflection on technology and decisive confrontation with it must happen in a realm that is, on the one hand, akin to the essence of technology and, on the other, fundamentally different from it".[9] For Heidegger, art was such a realm. Any new creation should consequently take place not only through technical skill, but also through *poiêsis* and the "artistic and poetical bringing into appearance and concrete imagery".[10] And this takes us back to Semper's knot, whose rational suitability also contains the germ of art and of poetry.

§9 __ Brick on brick. "Architecture starts when you carefully put two bricks together." It is the use of the word "carefully" that makes Mies van der Rohe's classical dictum important. It is easy to stack bricks. But through *careful* stacking, based on equal amounts of technical insight and artistic impulse, brickwork is given a touch of the splendour that makes it more than just a technical accomplishment of a given function. The codes of the bond are the knots with which the brickwork becomes an organic form. The "brick knot" is the tectonic basic component that gives the concrete physical creation a touch of cultural history's mysterious glamour. When brickwork is robbed of its flexibility through the use of cement mortar, and vertical dilatation joints must be cut through it as a result, 10,000 years of continuous cultural history are interrupted at the same time. Respect for the "tectonics of dressing", as the German architect Udo Garritzmann calls it, is one of the great architectural challenges of our time.[11]

**Brickwork is Earthwork**

§10 __ The cave. Semantically, bricks can be closely linked to protection, to the cave. The cave is part of the earth that is hollowed out in order to create a new "crust" to provide shelter. Clayey soil – and the bricks that are fired from

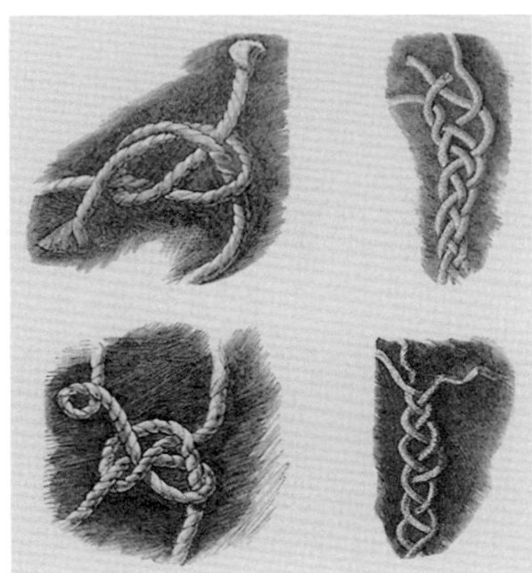

Top: Weaver in Peru, her infant patiently watching.
Centre: Knots and knittings, illustrations from Gottfried Semper's opus magnum *Der Stil* (1860).
© Semper *Der Stil*
Bottom: Student work, The Royal Danish Academy of Fine Arts, School of Architecture, Copenhagen (2009).
© Thomas Bo Jensen

it – is consequently a symbol of security that is intimately bound up with man's existence and relationship to the world, or with our *ability* to dwell, as Heidegger would say. "Only if we are capable of dwelling, only then can we build", he wrote, referring to the Old High German word for building, *buan,* which also means to dwell.[12] In other words, at the moment you take a lump of clay from the earth and hold it in your hand, it reminds you of dwelling, through the hollowing out of the earth, and of building, through its materiality. In this respect, clay is an existential element.

§11 __ Layer upon layer. In 1949, the German architect Rudolf Schwarz published *Von der Bebauung der Erde*. It is a very curious book, which deals with the close relationship that Schwarz found between architecture's constructions and the earth's geological structure. "We forget", he wrote, "that long before man's architecture existed, there was the earth's architecture, which is still being created".[13] For example, Schwarz compared the earth's sedimentary deposits with brickwork's alternation of bricks and joints. He found that brickwork's layered structure, built up course by course, was based on the same logic with which sediments have been deposited layer upon layer over millions of years. For Schwarz, brickwork consequently represents a kind of continuation of the earth's geological processes. His point was that when we build houses and cities, we are basically building upon what already exists.[14] This idea is especially relevant when we consider the brick, which over the millennia has been made of raw materials taken specifically from the earth's sedimentary deposits. But it is also a question of a more fundamental existential condition that an architect faces when he builds. Architecture is neither a beginning nor an end to something; it is a continuation – a continuation of a series of given conditions that are in constant flux, either slowly, like the formation and erosion of mountains, or somewhat more quickly, like the overlapping layers of cultural history. Brickwork, which can survive over millennia and in time be expanded and transformed for another use, can be seen as a metaphor of these ongoing processes.

§12 __ A section through matter and time. Rudolf Schwarz was especially interested in the cyclical formation of mountains. The sediments that come from the erosion of mountains are deposited layer upon layer in rivers and on the seabed, the oldest at the bottom and the youngest at the top. This stratification is also reflected when sedimentary layers are petrified and become mountains as the seabed rises. But if we look at a vertical wall of rock or a cross-section of a sedimentary massif, the layers are rarely horizontal. Through the enormous inner forces caused by shifts in the tectonic plates, they are in fact pressed upwards and deformed to such an extent that the layers run diagonally or vertically or take on a folded character. Schwarz pointed out that at times, the older layers fold around the younger ones and as a result turn chronology upside down, or else glide away from one another, and in doing so are randomly

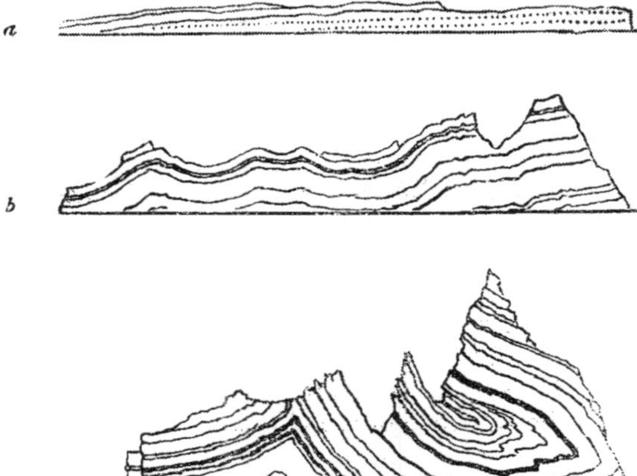


Fig. 16.

combined. “The section cuts through the building mass of matter and time,” he wrote, “so that what does not belong together is joined together, and what belongs together is pushed apart”.[15] Since the theory of plate tectonics did not gain ground until the 1960s, Schwarz was obviously unaware of the dual role that tectonics has assumed as a basic concept for the structural processes of both architecture and the earth. What makes it all the more interesting is that he used the German concept Architektonik when he described the way in which nature itself builds walls, columns and stairs. For Schwarz, these elements were not human discoveries; they were among nature’s building blocks, which human beings adopted and developed. “The earth constantly builds further upon its stairs and man uses this form... He ‘re-masons’ the mountains’ slopes.”[16]

§13 __ Ruskin. John Ruskin had had similar thoughts a century before when he travelled as a young man in William Turner’s footsteps to Switzerland to study the geological principles of form of the Alps. Ruskin had actually intended to study Turner’s motifs and painting techniques, but his fieldwork gradually developed into a full scientific study in which mountain formations and their analogies to architecture took up as much space as the study of art. Ruskin made a number of ink drawings that were used in the fourth volume of his mammoth *Modern Painters*. The chapter “The Sculpture of Mountains” features a series of drawings of sedimentary formations that point to the same characteristics that fascinated Schwarz. In Ruskin’s day, there was very limited knowledge of the age and formation processes of mountain ranges. Ruskin emphasised that he based his observations solely on *presumptions*.[17] For example, he described how the stratified structures of the mountains always “lie in waves” and must consequently have been “caused by convulsive movements of the earth’s surface”. He described how mountains could demonstrate “contortions”, “fractures” or breaks that had taken place with a “sharp crash”. And he noted that the effects must have been exerted in a single contemporaneous movement since these mechanisms always seem to be interrelated.[18] Ruskin was consequently on the right track, although the theory of continental drift was not formulated until around 1915 by the German polar researcher and geophysicist Alfred Wegener, and not accepted by the geoscience community until 1966. This was when the Canadian geologist John Tuzo Wilson had presented evidence of the existence of the supercontinent Pangea (yet another one of Wegener’s controversial ideas) – and consequently also of the theory of tectonic plates as the main “driver” for the earth’s geological processes of change.[19]

§14 __ Kirkeby. But all this was actually not Ruskin’s focus. Rather than seeking geoscientific truth, he was concerned with how natural contortions, fractures and crashes can work as a driving force for the artistic process of creation. In Per Kirkeby’s view, this is what makes Ruskin interesting to this very day. Kirkeby took his degree in geology at the University of Copenhagen in

Top left: Two sketches of sedimentary mountains, Rudolf Schwarz, *Von der Bebauung der Erde* (1949).

Bottom left: Three sketches of sedimentary mountains, John Ruskin, *Modern Painters* Vol. 5 (1856).

Right: Three diagrammatic drawings by Rudolf Schwarz, *Von der Bebauung der Erde* (1949). The upper drawing shows what Schwarz defines as nature’s own columns; the centre drawing shows the sedimentary structure of the earth (“course by course”); bottom drawing shows what Schwarz sees as nature’s own stairs.

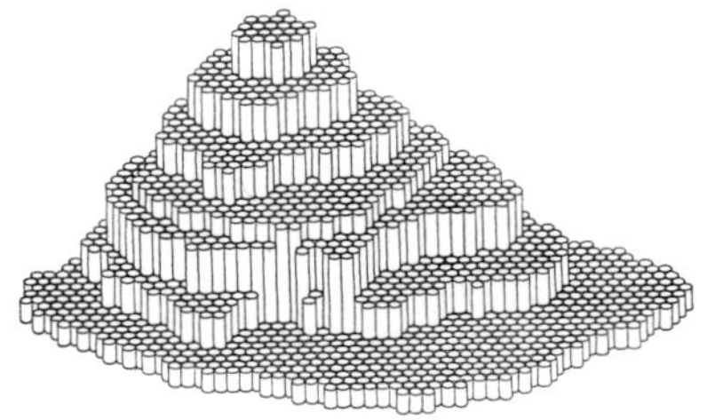

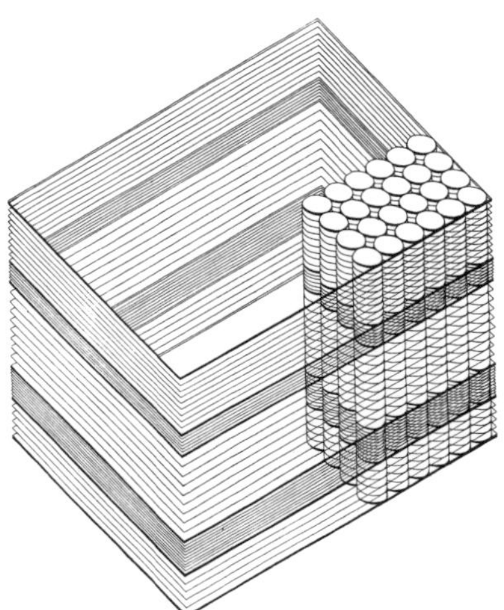

1964, majoring in arctic geology. He made several study trips to Greenland as a student and took a keen interest in the field's controversies about continental drift and tectonic plates. He has kept this scientific interest alive throughout his career as an artist, with the fells of Greenland serving as a recurrent "secret reservoir" of inspiration.[20] For Kirkeby, the theory of continental drift and the geological concepts associated with it (plate tectonics, faults, folding, sliding, etc.) are first and foremost a poetic translation of a number of scientifically proven processes whose resulting forms can be read directly. These concepts are joined by the geological figures that are associated with the processes of erosion, for example the talus cone, which is used in much of his visual art.

The concept of tectonics is especially important to Kirkeby. It is a kind of driving force in his art, including his work with "tectonic bricks", as he likes to call them.[21] For Kirkeby, tectonics is a kind of petrified movement, which has found its logical form as a result of inner pressure. This inner pressure comes from the material itself. For the brick, it lies in its format and the constraints that it imposes: "limitation as a determinant", rather than freedom to do anything at all.[22] In Kirkeby's brick buildings, (poetic) fundamental geological principles leave their mark directly on the brickwork. Faults manifest themselves as shifts in the wall's plane; folds are found as vertical, shadowy slits, and slides as overlapping walls that thrust themselves in front of one another. Kirkeby's point is that these forms were derived not as a direct result of an artist's interest in geology, but as a result of the brick's inherent characteristics. Geological principles of form nevertheless make it easier to organise the architecture and can even provide original solutions to static problems and artistic forms for the dilatation joints that otherwise seem to challenge the brickwork's cohesion. But for Kirkeby, the opposite is true. For him, the need for dilatation joints is a confirmation of architecture's kinship with the fundamental principles of geology.

The tectonic kinship between architecture and nature is especially conspicuous in Kirkeby's work. In a way, his architecture is a realisation of Rudolf Schwarz's dream of approaching architecture as if it were woven into the "world order" or part of "a world theatre of creation and transience".[23] Perhaps this is why his architecture appears to be arid and angular, at times empty, curious or contradictory. This is not architecture that aims at balance, where the individual parts are conceived as forming a harmonious whole, something that an architect would typically wish to achieve. In a way, Kirkeby's architecture is uncultivated, like nature's abrupt transitions and crashes.

**Brickwork is Recollection**

§15. "Why use brick?" This is a question that Per Kirkeby often asks himself. He closes his eyes and tries to find the answer. Each time, he conjures up his childhood in Copenhagen's Bispebjerg quarter – the Grundtvig Church and the surrounding area, with endless rows of modest brick apartment houses from the 1930s and 1940s. "It is the atmosphere, the brick's accumulation of memory and history", he notes.[24] When he began to work with bricks in 1965,

Top left: Vesthimmerlands Museum, Aars, Per Kirkeby and Jens Bertelsen (1999). © Ole Meyer
Centre left: Smoothed brickwork, Grundtvigs Church, Copenhagen, P.V. Jensen-Klint (1921-1940). © Ole Meyer
Bottom left: Scratched brickwork, old naval base, Holmen, Copenhagen. © Thomas Bo Jensen
Right: Side nave of Grundtvigs Church, Copenhagen, P.V. Jensen-Klint (1921-1940). © Ole Meyer

they were literally smuggled in through the back door, to the amazement of his avant-gardist colleagues. "Even though one looked innocent and followed all the rules", the bricks brought this whole complex of symbol-charged significance and aesthetic elements such as color, light and texture into minimalism's stringent exorcist project. This inevitable baggage is the reason why brick is often suppressed when new waves break in architecture. It is a bit too intrusive to signal innovation. It is too loaded, too murky - and a bit too sluggish. This combination is precisely what makes bricks so fascinating to Kirkeby. "In a manner of speaking, they are there before we see the art."[25]

§16 __ The brick is materialised time. Not many years have to pass before dust and soot settle on the brick's textured ceramic surface. Not in an ugly way, like dirt and algae on façade plates that must be regularly cleaned to keep a building looking new and fresh. It is different with bricks. Their patina makes them more vibrant over time. If carried out painstakingly, brickwork will slowly become imbedded in a culture. The same thing can naturally be said of other heavy materials such as concrete and stone, but brick is different. Because of the fired brick's softness, it more easily shows signs of wear. This is something that P. V. Jensen-Klint exploited when he asked his masons to smooth every brick that was used in the Grundtvig Church by hand before it was laid in place. The traces of the masons' working hands were inscribed for all time in each and every brick - an almost breathtaking thought. And this is what generation after generation of young naval recruits knew when they scratched their initials in the yellow bricks at the Holmen naval base in Copenhagen. Now the Royal Danish Academy of Fine Arts has taken over the buildings, but the voices from the past still remain - thanks to the softness of the bricks.

§17 __ Philosophical clay. Richard Sennett considers clay to be "that most philosophical of materials".[26] He notes that because of its great wealth of nuances, brick has often been attributed with anthropomorphic qualities. Brick can evoke human metaphors, for example, when heavily patinated brickwork is described as "an old man's weathered face", or is credited with human characteristics such as "warmth", "honesty" or "modesty". The use of these metaphors is meaningful to Sennett since they can help heighten our consciousness of materials "and in this way [lead us] to think about their value".[27] Sennett then considers the possibility of dropping all the lyricism and just treating clay as a useful material that is necessary for providing shelter, etc. "But in this utilitarian spirit we would eliminate most of what has made this substance culturally consequent", he emphasises.[28]

### Future Brickwork

§18 __ A large part of what has made brick culturally consequent is being challenged today. The individual bricks still have their texture, colour and shimmer, but these features are in fact the only positive things left. This might

sound like just another conservative lament about cultural loss. But if the baby that we throw out with the bathwater has precisely the qualities that can promote the development of a more sustainable building culture, then it is no longer a lament that can be shrugged off as sentimental nonsense. Cultural robustness, durability, minimal maintenance, technical flexibility, thermic qualities, etc. – all these are just some of the many potential advantages of brickwork.

The partnership of the hand and the intellect is inseparably joined in the cultural history of clay and brick. Although robots have been invented, it is still the hand's programming abilities and the horizons of the intellect that determine the final results. Contemporary brickwork must raise its level of ambition so we can ensure that our own age will also leave traces of lasting value. If we understand our work as part of a bridge that links generations of mankind, then clay, and the experiences that are associated with its processing and use, might be our best guide – in developing the brickwork of the future.

1 __ Lars Morell, *The artist as polyhistor* (Aarhus: Aarhus University Press, 2005), p. 156.

2 __ Gottfried Semper, *Style in the Technical and Tectonic Arts, or, Practical Aesthetics* (Los Angeles: Getty Publications, 2004), p. 218 ff. English edition of *Der Stil in den tektonischen Künsten, oder praktische Ästhetik*, vol. 1, *Die textile Kunst für sich bearbeitet und in Beziehung zur Baukunst* (Frankfurt am Main, 1860).

3 __ Semper, *Der Stil*, vol. 1, Prolegomenon.

4 __ Mari Hvattum, "Den nakne sannhet – et essay om ornamentik og arkitektur," *Norsk Byggekunst* 4 (2005), p. 17.

5 __ Semper, *Style*, p. 218 ff.

6 __ Martin Heidegger, *Basic writings* (New York: Harper Collins Publishers, 1977, 1993), p. 317.

7 __ Heidegger, *Basic writings*, p. 318.

8 __ Dan Zahavi, *Martin Heidegger, Spørgsmålet om teknikken og andre skrifter* (Copenhagen: Gyldendal, 1999), p. 27.

9 __ Zahavi, *Martin Heidegger*, p. 31.

10 __ Heidegger, *Basic writings*, p. 317.

11 __ Udo Garritzmann, "*Murværkets tektonik – en tektonik om beklædningen eller den bærende konstruktion*," *Arkitekten* 4 (2013), p. 28 ff.

12 __ Martin Heidegger, "Tænke bygge bo," *Grid* 18-19 (2000), p. 2.

13 __ Rudolf Schwarz, *Von der Bebauung der Erde* (Salzburg/Munich: Verlag Anton Pustet, 2006, reprint from 1949), p. 22.

14 __ Schwarz, *Von der Bebauung der Erde*, pp. 22-24.

15 __ Schwarz, *Von der Bebauung der Erde*, p. 30.

16 __ Schwarz, *Von der Bebauung der Erde*, p. 35.

17 __ John Ruskin, *Modern Painters*, vol. IV, *Of Mountain Beauty* (London: Smith, Elder and Company, 1856), p. 152. "I describe facts or semblances, not operations. I say '*seem* to have been,' not '*have* been.' I say '*are* bent'; I do not say '*have been* bent'."

18 __ Ruskin, *Modern Painters*, vol. IV, pp. 153, 158-159.

19 __ Ron Redfern, *Origins: The evolution of continents, oceans and life* (London: Castell & Co, 2000), p. 77.

20 __ Per Kirkeby, Ane Hejlskov Larsen, and Erik Steffensen, *Per Kirkeby og Grønland. Det hemmelige reservoir* (Charlottenlund: Ordrupgaard, 2012), p. 11.

21 __ Per Kirkeby, *Bravura* (Copenhagen: Borgen, 1981), p. 44.

22 __ Per Kirkeby, *Håndbog* (Copenhagen: Borgen, 1991), pp. 45-46.

23 __ Schwarz, *Von der Bebauung der Erde*, p. 4.

24 __ Per Kirkeby, *Fisters klumme* (Copenhagen: Borgen, 1995), pp. 96-97.

25 __ Kirkeby, *Fisters klumme*, p. 112.

26 __ Richard Sennett, *The Craftsman* (London: Penguin Books, 2009), p. 125.

27 __ Sennett, *The Craftsman*, p. 144.

28 __ Sennett, *The Craftsman*, p. 151.

**Thomas Bo Jensen**
is an associate professor at The Royal Danish Academy of Fine Arts, School of Architecture, in Copenhagen. In 2002 he defended his PhD thesis "The ornamental will of the brick". Since then he has been teaching and doing research within the field of tectonics with a special focus on brickwork and masonry. Besides working in this field he has established himself as the author of two large monographs, one focusing on the work and life of the Danish architect P.V. Jensen-Klint (published in 2006/2009) and the other describing the practice of Johannes and Inger Exner (published in 2012). Thomas Bo Jensen currently chairs the Institute of Architectural Technology at KADK.

Per Kirkeby, watercolour drawing (assumed 1997). Cliff's projecting a folded wall.

Johan Celsing

# Gravity and Clemency

The tectonic aspect in the discipline of architecture has informed my work in various, though not very clear, ways since I became aware of it as a concept. I have looked upon it as a strategy for designing buildings with a material intelligibility, rather than as one focused on visual appearance.

I have drawn inspiration from what I have read by Gottfried Semper and others on this subject. The ideas presented by Semper in the text *Die vier Elemente der Baukunst*[1] and the concept of "Bekleidungstheorie"[2] have had some influence. Like Semper, the buildings and writings of Adolf Loos have also meant a lot to me. At the same time, another, quite different aspect has become increasingly important in my work: the writings of Leon Battista Alberti, in particular the ten books on the art of building, *De re aedificatoria,*[3] which contain his discussions about *decorum* (the appropriate) in architecture and in life. I find this important philosophical concept, which has developed since Aristotelian times, very useful even today.

However, having said this, I have to admit that I myself would certainly not classify my work as tectonic or theoretical. My work is informed by so many aspects, some developed over a long time through experiences in earlier projects, some from life or literature, many of them vague, personal or intuitive. I also make a lot of decisions on the building site itself, because of the immediate need for solving a problem together with the craftsmen.

This text will focus on the recently inaugurated Church at Årsta with comments on two other projects I developed during the same period. All three projects are in brick, but with varied treatments. The varied tectonic solutions are presented as a way of widening the perspective: ranging from the gravity of the church's masonry to lighter, even wilful solutions in the brickwork of the other projects. In the presentation of the projects I will, in addition to the tectonic approach, describe the overall intentions that form the basis of the work: the atmosphere; how the work relates to its location; the wishes of the client; the use; and its current, or expected, social role.

New Crematorium Woodland Cemetery, Stockholm © Johan Celsing Arkitektkontor

Exterior, façade to the south (private house Lidingö) © Ioana Marinescu
Detail of glazed partitions to Pool Room. © Ioana Marinescu
Pool Room, in white concrete. © Ioana Marinescu

### House at Lidingö

Very near the Millesgården Art Gallery, which I designed a decade ago, we recently completed this private house. The client had three definite wishes, which were all different from what their previous house could provide: an indoor swimming pool, a terrace overlooking the archipelago outside Stockholm, and a flat roof. This building was the first of the three projects mentioned in this text and was to some extent a testing ground for many of the solutions, which were further elaborated in the later two.

The house is a compact volume when compared to the large villas in this well-to-do Stockholm suburb. Like the other projects, it has a brick exterior, which is contrasted on the inside with surfaces of lighter and more varied colour schemes.

Here the brickwork is a cladding on top of an in-situ cast concrete structure. The fact that the exterior brickwork is not load-bearing is evident, at least to the trained eye. Laid on the long side, the bricks are arranged so as to create vertical seams at the edges of each brick. This arrangement is meant to give a distinct but subtle pattern to the elevations. With the location in a sloping terrain, the position of the pool space is at the lowest level of the house, where wide, sliding, glazed partitions open to the west, while a vertical light well opens to the sky at the eastern side of the house.

The pool space interior is made entirely of white concrete, using white Danish cement from Aalborg, with ballast of white dolomite. The white concrete covers all the surfaces: from the walls and the ceiling to the floors and even the pool itself. My intention, which was quite a challenge for the contractors, was that the surfaces should not be treated after the casting was finished. Therefore, utmost attention was given to the plywood formwork and its edges, so that the resulting surface would have a soft, almost sensual quality, in which the weight and the production process could still be perceived. The living spaces of the house were designed to give a gentle character in terms of materials and details. Massive, lye-treated, wooden floors lend a soft character to the rooms, while occasional variations of colour add to the ambience.

### Årsta Church

The 1940s suburb of Årsta is located close to the southern part of Stockholm's city centre. The church is located on a rocky hill overlooking the main square, which houses several public buildings designed by the distinguished Swedish architects Erik and Tore Ahlsén (1901–1988 and 1906–1991).

This project started in 2006 with an invited competition. Prior to this there had been plans for a church to be built, ever since the area was developed after the Second World War. In 1952 a bell tower was built on the site. In 1968 a parish building was built, and the parish hall served as a church until last year. Our proposal for the competition was a freestanding, cubic volume in red and white. After we won the competition, the congregation urged me to connect

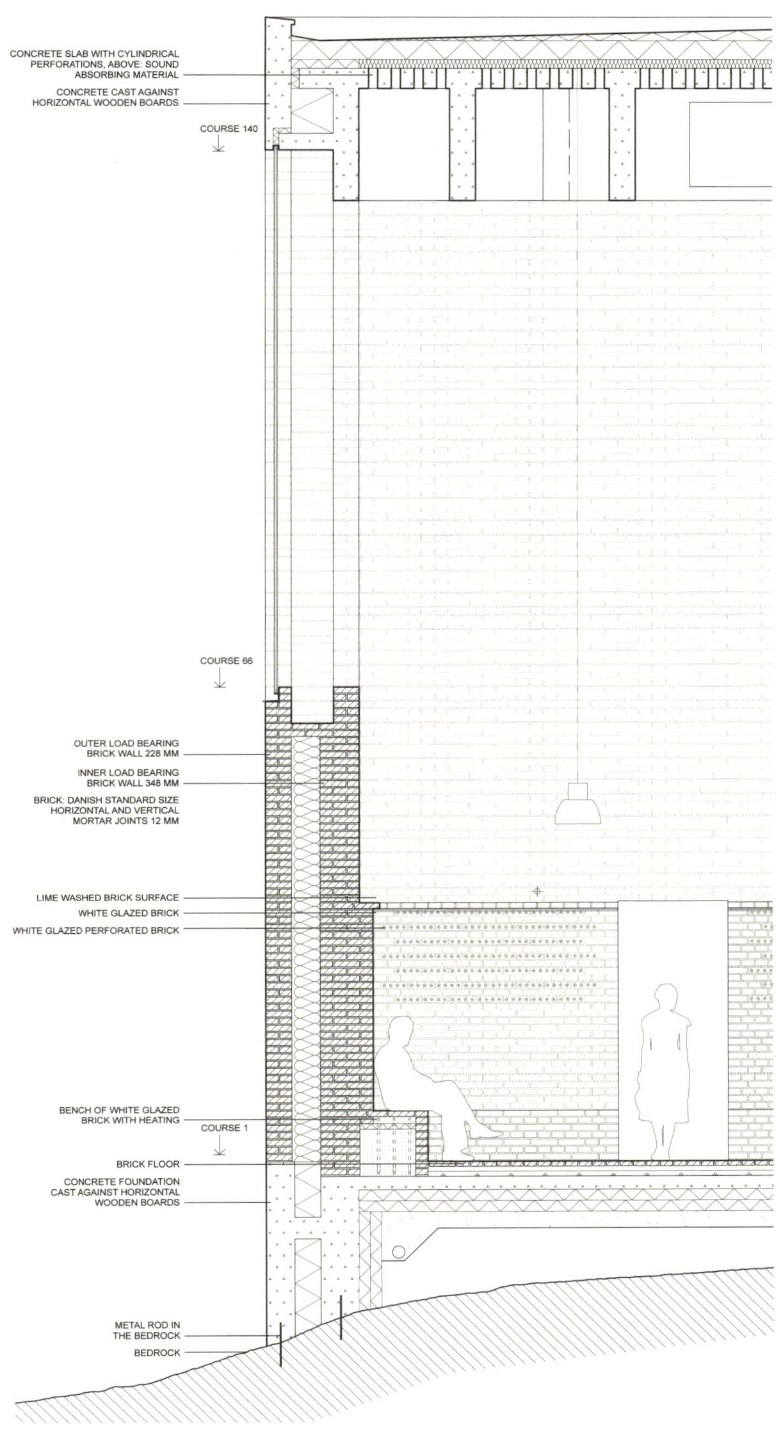


Exterior from the southwest © Ioana Marinescu

Principal section, massive brick wall, 88 cm, with cavity for thermal insulation and interior glazed brick surface . © Johan Celsing Arkitektkontor

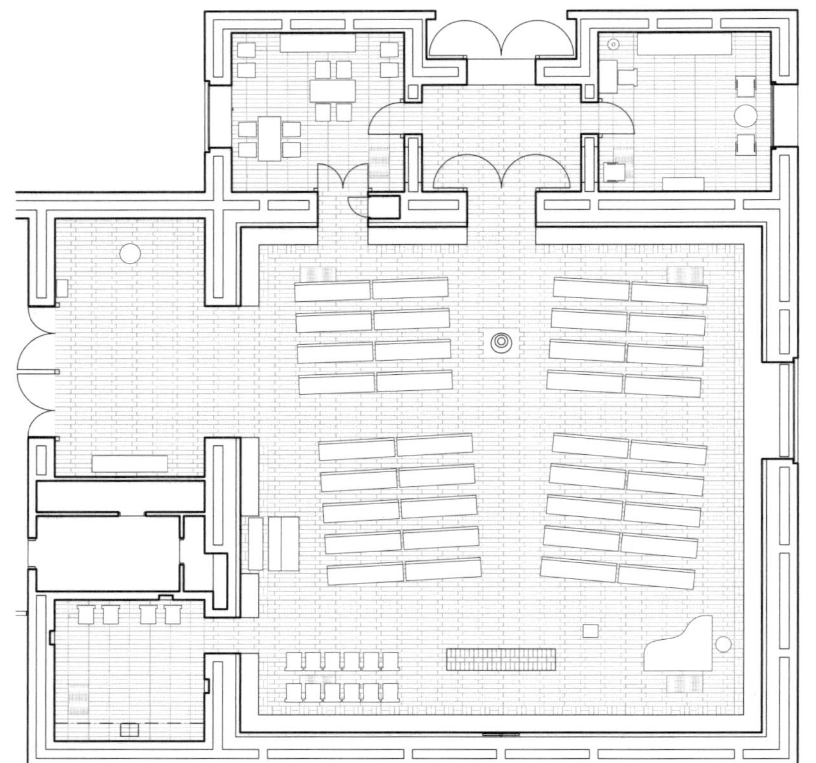

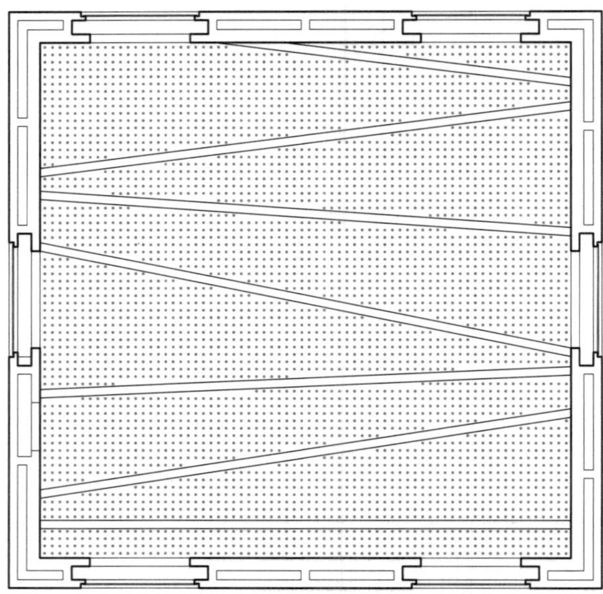

Plan (of the Church). © Johan Celsing Arkitektkontor

Ceiling plan , showing in-situ cast concrete beams and horizontal surface with perforations and positions of main windows with lateral niches. © Johan Celsing Arkitektkontor

Sketch of section, showing openings, profiling and bench along the outer walls. © Johan Celsing Arkitektkontor.

Interior wih outer wall with continous bench and surface of white glazed brick. © Ioana Marinescu

Interior of Church with glazed brick and continuous bench. © Ioana Marinescu

Western Entry with view to the Sacristy © Ioana Marinescu
Interior of Children's Chapel with surface of green glazed brick. © Ioana Marinescu
Detail of baptismal font. © Johan Celsing Arkitektkontor

the church to the existing parish building, which was low and constructed in dark brick. Its greatest asset was a generous atrium courtyard. The request was very reasonable, since several additional spaces, such as cloakroom, toilets etc., were already located in the existing building. However, the merging of the two buildings made the red-white exterior redundant. The merging of the two buildings was the reason for giving the church brick as its major tectonic component. Whereas the parish building only had brick as cladding, the church has brick for the load-bearing structure as well as for the surfaces for both interior and façades. Technically the building is of massive masonry. However, due to Swedish regulations, the walls are built as cavity walls with thermal insulation. As is shown in the section, the bedrock under the church is used as a foundation, with metal rods drilled into the rock as armature for the concrete foundation.

Early studies for the gestation of the church included numerous elaborations of an intricate ceiling, where celestial light (inspired by the churches of the Baroque) would enter through hidden skylights. In the search for what seemed most appropriate, I decided to skip these designs. In the process of developing the concept, it became evident that the most important aspect of the architecture should be to bring concentration and dignity to the primary rituals of the space, such as baptism, divine services, funerals and weddings.

The floor area of the church is only 13 by 13 metres, while the design of the church was inspired by the idea of the standing congregation of an Orthodox church. The pews, which accentuate the profile of the walls, would then provide seating for the elderly.

For their tactile and glimmering qualities, the lower part of the walls was built using white-glazed bricks. Further up, the walls are painted with several layers of traditional lime wash. The walls are massive and load-bearing: 88 cm thick, with cavities for insulation, but without expansion joints. To articulate the quality of the light entering the church, the windows have lateral niches of almost man-sized dimensions. Above the large windows (280 x 500 cm) there is no brickwork, but a continuous, concrete cornice, cast with the use of horizontal wooden boards.

The ceiling is made of in-situ cast concrete. The horizontal surfaces have perforations (diameter 75 mm) set 200 mm apart. Their prime function is acoustic: they lower the reverb in the church. Spatially they are intended to enhance the surface of the concrete with their repetitive pattern. The detailing of the walls, ceiling and furniture is intended to complement the elementary space with a calm but palpable sense of care. The sacristy has walls with red-glazed bricks. The children's chapel has green-glazed bricks and chubby chairs of an ambiguous size and elementary character, which to a child may seem almost possible to fabricate by yourself.

The position of the altar and font on the east-western axis is intended not only to provide a sense of direction in the square space, but also to mark the liturgical function even when the church is used for other activities such as music or recitations. The Swedish Brännlycke marble font is positioned on a large slab of the same marble, carefully tessellated into the brickwork pattern of the floor. The western entry faces a new outdoor entry area, which overlooks the neighbourhood. Here, after the service, the congregation, mourners or wedding guests can gather at the foot of the bell tower. The bell tower also has a strong visual presence at the main entrance, located where the church joins the parish building. When connecting the church with the existing building, the objective was to cast the two parts into a new entity. When merging rooms such as the existing entrance lobby, which was doubled in size, we made a particular effort to maintain integrity in terms of new and old surfaces and characters. In the new part of this particular room the in-situ cast ceiling was cast on a formwork of wooden boards, of the same kind and dimensions as the whitewashed suspended ceiling in the old part, which was left unaltered.

In the design I had no ambition to distinguish the old from the new or to emphasise the edges where they meet. This attitude also prevails on the exterior. However, at a close range or for the attentive visitor, it is evident that the new church was built using very different bricks and with quite different objectives from those of the original parish building.

**The New Crematorium at the Woodland Cemetery**

We are currently constructing a new crematorium at the Woodland Cemetery in Stockholm. The cemetery was designed by Sigurd Lewerentz and Erik Gunnar Asplund after they won a competition in 1916 with a joint project, and was subsequently elaborated over a period of 25 years. The present project started as an international competition by invitation in 2009. The new building will function as a crematorium with spaces where mourners may gather to take part in the cremation rituals. The location is in a densely forested area of the cemetery and in the immediate vicinity of the existing crematorium. It was chosen by the authorities so as not to disturb the quality of the World Heritage buildings and the existing landscape. As a response to the site and the brief, the building is a compact volume organised in response to the undulating terrain. The motto for the competition was "A stone in the forest". Our project is set back into the woods and is reached by a rising driveway and path. The intention is that mourners walking up to the crematorium on the path between the trees may find time and space to prepare themselves for what will take place in the building. For visitors the entry is announced and sheltered by a wide canopy, under which groups of mourners may gather.

The entire exterior, façades and roof are in brick, built on top of an in-situ cast concrete structure. The hard-burned Danish bricks measure 520 x 108 x 34 mm. The insulation behind the skin of brick is the inorganic material perlite. The exterior brickwork has no regulated pattern. The roof, which is technically a terrace, has the bricks laid on the flat side. The details are shown in the wall/roof detail drawing. Visually, the exterior is held together by the large asymmetrical

Left: Furnace hall. © Ioana Marinescu
Right: Chapel. © Ioana Marinescu
Following pages: Canopy at Main Entrance. © Ioana Marinescu

roof, which is perforated by slits, recesses and windows, bringing light and air into the interior.

The interiors of the building are in contrast to the exterior in terms of shape, material and colours. All the major spaces are in all-white concrete. The actual quality of the formwork will be evident once the forms have been dismantled. When the forms are taken down, there will be no treatment of the white concrete surface.

The large Furnace Hall has a long wall entirely clad in perforated white bricks. This brilliant surface under the windows to the sky will reflect light into the hall, but also decrease the rumbling sound of the furnaces. Whereas the Furnace Hall ceiling follows the shape of the exterior roof, the Mourners' Room has a barrel-vaulted ceiling. This room is where the relatives may bid farewell to the deceased, who is in an open coffin, or where they may hold a ceremony when receiving the urn. This room has a narrow light slit at the top of the end wall, which has a white, glazed brick surface.

As a recess in the building block there is an atrium, where the light concrete interior meets the brick exterior. This is an outdoor space where the staff may gather for breaks. Continuous glazing separates the surrounding white corridors from the atrium, where the brick façades frame the pines and the sky above.

The major public space, however, is under the canopy. Here, next to the wooded landscape, the surfaces are all in brick in different positions: on the façades; as paving; and cast into the ceiling. The structure here clearly has its own intentional use of brick. There are also brick piers and a massive granite column the size of a tree trunk, which supports the canopy. Under this roof mourners may gather and, if present during the cremation, may rest and seek comfort in the forest and its sounds.

One could add generally that the design of these buildings has been inspired by the formal qualities and content of the lyrical works of poets such as W.H. Auden (1907–1973) and Inger Christensen (1933–2009).

> "Blessed be all the metrical rules
> That forbid automatic responses
> Force us to have second thoughts
> Free from the fetters of Self."[4]
>
> *W.H. Auden*

**Johan Celsing**
founded the Swedish architectural practice Johan Celsing Arkitektkontor in 1984. Over the years the Stockholm-based firm has gained international recognition for buildings such as: The New Art Gallery for Millesgården Museum, Stockholm; the extension of Skissernas museum, Lund; and Årsta Church, Stockholm. Besides being a practising architect, Johan Celsing plays an important role in the discussion of architecture within a Nordic context. In 2008 he was appointed professor at Kungliga Tekniska Högskolan in Stockholm.

1 __ Gottfried Semper, *Die vier Elemente der Baukunst* (Braunschweig: Druck und Verlag von Friedrich Vieweg und Sohn, 1851)

2 __ Gottfried Semper, *Der Stil in den technischen und tektonischen Künsten: Bd. Die textile Kunst für sich betrachtet und in Beziehung zur Baukunst* (Frankfurt a. M.: Verlag für Kunst und Wissenschaft, 1860)

3 __ *De re aedificatoria* was written by Alberti between 1443 and 1452. An English translation can be found in: Leon Battista Alberti, *On the Art of Building in Ten Books*, trans. Joseph Rykwert, Neil Leach and Robert Tavernor (Cambridge, Massachusetts: The MIT Press, 1988)

4 __ W. H. Auden, *Epistle to a Godson* (London: Faber & Faber, 1972)

Atrium © Johan Celsing Arkitektkontor

Jonathan Hale

# Cognitive Tectonics: From the Prehuman to the Posthuman

Writing in 1993 on the relations between technology, language and cognition, the anthropologist Tim Ingold provided what appeared to be a perfectly clear and succinct definition of the tool as a "prosthetic" extension of the body:

*"A tool, in the most general sense, is an object that extends the capacity of an agent to operate within a given environment."*[1]

On closer inspection, it could be argued that this quotation actually assumes what it sets out to explain: that is, that we already know what constitutes an "agent" in this context, and that we can therefore speak of the tool as a simple linear extension of an agent's ability. In fact, if we consider this question within the "long duration" of the evolutionary emergence of the modern human being, it may be more accurate to say that the tool, in reality, came first. Or, at the very least, it could be argued that technology is mutually co-implicated in the emergence of human agency itself.

In order to explore this apparently circular relationship between the human and the technological, in what follows I will discuss some examples of the ways in which we engage with technologies on a day-to-day level, and how the process of "incorporation" (literally, absorbing into our body-image or, more accurately, our body-schema) entails a number of important cognitive consequences. In the final part of the chapter I will also try to outline what I think this might mean for the continuing relevance of tectonic articulation in architecture: for example, in the creation of engaging and richly layered environments which contain visible traces of both construction and occupation, spaces which invite engagement with both the bodies and minds of future building users.

### Technology and Embodiment

The classic example of the incorporation of the tool into an extended body-schema is that of a blind person learning to navigate with the aid of a white cane. I take this illustration from the writings of the French philosopher Maurice Merleau-Ponty, who described it in his major work *The Phenomenology of*

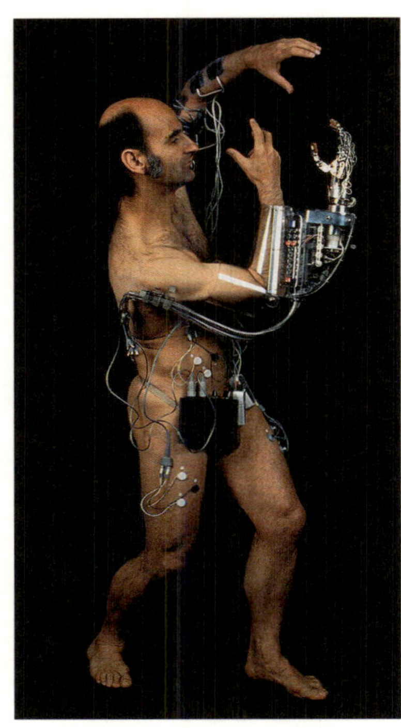

Stelarc, *Third hand*, Tokyo, Yokohama, Nagoya 1980. Photographer: Toshifumi Ike. © Stelarc / T. Ike.

*Perception* (1945).[2] Through a gradual process of exploration and experiment the sensitive surface of the hand is effectively stretched out towards the tip of the cane: information is gathered as the cane reaches out, and by experiencing the textures of touch and sound an environment begins to be revealed. With skilful use the cane effectively "disappears" from view, as Merleau-Ponty suggests it ceases to be an object that we perceive in itself, and instead becomes a "medium" through which we experience the world, just like the body itself. The use of the cane is gradually sedimented into a behavioural and therefore perceptual routine: it becomes part of the repertoire of bodily skills and abilities which we use every day to navigate our way through familiar and not-so-familiar environments.

Another more dramatic illustration of this flexible boundary between brain, body and world can be seen in the experiment carried out by the Australian performance artist Stelarc, who added a prosthetic "third hand" to his own biological body[3] (Massumi, 1998: 336). The hand is controlled by nerve impulses picked up from surface electrodes attached to his upper thigh and abdominal area. While the device took some time to learn how to operate (basically by a process of trial-and-error experiment), with practice it could be quite precisely controlled, independently of the artist's two biological hands.

This example also reminds us of the fact that, from birth onwards, we have all passed through a similar process of exploratory bodily "training", swinging our limbs about more or less wildly until we gradually learned how to control and apply them with some precision.

The idea that technical extensions of the body can become intrinsic to our individual "sense of self" is also suggested by the philosophers Andy Clark and David Chalmers in their 1998 essay "The Extended Mind".[4] They describe how we commonly rely on various technical props and supports to help us deal with everyday mental tasks: from notepads and pencils for writing down ideas to electronic calculators and digital search engines for retrieving and manipulating useful information. The all-too-familiar misfortune of losing a wallet or a mobile phone also reminds us how distressing it is to be denied access to what can suddenly seem like a vital organ. Robbed of our taken-for-granted ability to make phone calls, look up addresses, check diary entries and access the Internet, it is easy to feel that we are not quite the complete person which we previously assumed we were.

As if to dramatise this situation, the French philosopher Bernard Stiegler, in his recent book *Technics and Time*, even goes as far as to say that, far from being simply "optional extras", these technological extensions which we routinely incorporate into our extended body-schema should be seen as a fundamental part of what it is to be a human being.[5] In the next section I will explore this idea within an evolutionary framework, drawing an analogy between the ontogenetic processes we have just been considering (the development of the embodied individual enhanced by various technical extensions) and the longer timescale of the phylogenetic process by which the human species itself can be seen to have emerged. To do this, I will apply a model of "circular causality": the idea that a kind of feedback loop between technical development and biological mutation helped to steer the course of human evolution. Or, in relation to architecture, as Winston Churchill once famously said:

*"We make our buildings and thereafter our buildings make us."*

My first piece of evidence is taken from a recent book called *The Prehistory of the Mind* by the cognitive archaeologist Steven Mithen. In it he shows a timeline for the development of early hominid species, showing increases in average brain size over the last 4 million years.[6] The key points are the two major periods of significant brain enlargement, initially from about 2 million years ago, and then again from half a million to 200,000 years ago. In parallel with these developments archaeologists have also found evidence of the emergence of early stone tool technology, in the period from 2.6 million years ago up to 250,000 years ago, showing the increasing complexity of strategic planning involved in the transition from so-called Oldowan to the more advanced Late Acheulean tool-making processes.

Now, of course it is difficult to infer direct causality in one direction or the other: one might claim that bigger brains are the "cause" of more complex technology. Equally, I could try to claim that it is actually the other way round: the existence of more complex tool-making practices could be the selective pressure required to "cause" the preservation of genetic mutations, which happen to confer additional tool-making ability. What I actually want to claim here is simply that both these forces are interacting in a circular process of mutual support.

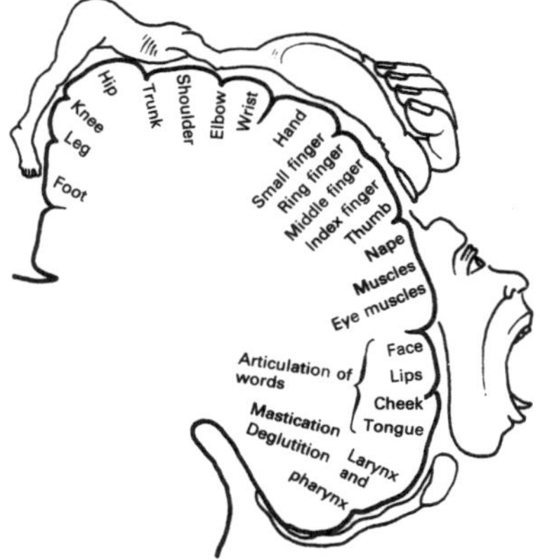


Leroi-Gourhan, *Cortical picture of voluntary motor function* (after Penfield and Rasmussen). Right hemisphere of human brain in cross-section. © MIT Press, 1993. P. 82.

Another important point worth noting about this evolutionary development is the relation between these early technical practices and the emergence of spoken language. While evidence for the existence of language is notoriously difficult to find, we can at least infer it from fossilised anatomical fragments. The increasing complexity of the vocal tract, as well as the size and shape of the skull, both imply the possibility of early human linguistic ability. There is also the circumstantial evidence of sophisticated social interaction suggested by the archaeological remains of complex communal settlements.

In his research from the 1960s onwards into the connection between gesture and speech, the paleoanthropologist André Leroi-Gourhan also found evidence of a further feedback loop between technical and linguistic ability. A key part of his evidence was based on the organisation of brain activity in the sensory-motor cortex, where the major areas devoted to control of the hands and the vocal apparatus are located in immediately adjacent areas. He supports his claims with a dramatic image taken from the work of neurologists Penfield and Rasmussen in the 1940s, showing where the major parts of the body's sensory-motor apparatus are processed within the brain.[7]

This diagram also emphasises the variation in the proportions of cortical space given over to the different parts of the body, with the largest areas devoted to those needing the most precise levels of control and articulation.

Leroi-Gourhan also partly bases his argument on the evolutionary shift from moving on all fours to walking upright, claiming that this innovation simultaneously liberated both the hands and the face for new uses. In place of their previous focus on ground-based movement and exploration, both can now be employed in new forms of communication. He further suggests that manual ability with tool-making might have provided the initial stimulus to the use of the hands to communicate, thus encouraging a refinement of a kind of proto-language of bodily posture into a set of more precisely articulated manual gestures. This idea has been further developed more recently in the work of the evolutionary psychologist Michael Corballis, in a book called *From Hand to Mouth* published in 2002,[8] as well as by Tim Ingold, whom I mentioned earlier, in an edited volume called *Tools, Language and Cognition in Human Evolution.*[9] In my view, Ingold makes the connection more convincingly, partly by the way he considers language itself as a form of technology: another kind of "tool" for reaching out beyond the body to make things happen in the world around us.

Leroi-Gourhan, for his part, accepts the speculative nature of this connection, given that the early historical traces of spoken language have clearly not been physically preserved. But if we consider the ways in which technical processes and procedures are visibly evident in the form of the tool itself, then perhaps this provides an example of how bodily communication can be "captured" and passed on from one generation to the next. The French philosopher Jacques Derrida, intrigued by the apparently inferior philosophical status of written, as opposed to spoken, forms of speech, was also inspired directly by Leroi-Gourhan to speculate on the evolutionary function which physical traces of human memory might actually have performed:

*"If the expression ventured by Leroi-Gourhan is accepted, one could speak of a 'liberation of memory,' of an exteriorization always already begun but always larger than the trace which, beginning from the elementary programs of so-called 'instinctive' behaviour up to the constitution of electronic card-indexes and reading machines, enlarges difference and the possibility of putting it in reserve: it at once and in the same movement constitutes and effaces so-called conscious subjectivity."*[10]

With the reference to "so-called instinctive behaviour" Derrida connects an original impulse towards mark-making with Merleau-Ponty's description of bodily skills and habits as our primordial means of grasping our place and finding our way within the world. That is, he implies that we should think of habitual patterns of behaviour as being our first means of capturing and passing on our acquired knowledge of the world.[11] Towards the end of the quotation Derrida also suggests that these forms of "exteriorization of memory", even as far back as the simple tools we have been discussing, gradually become both more elaborate and more durable, and perhaps also therefore mark the dawning of human self-consciousness itself – in other words, the beginning of human self-realisation which is, as he says, both "constituted and effaced" in the process which both preserves and objectifies the identity of the maker within the artefact, at the same time projecting it out into the world so it takes its place among countless other more or less "anonymous" objects.

For another way of looking at this, we could turn to the writing of the

> *"I would have objectified in my production my individuality and its peculiarity and thus both in my activity enjoyed an INDIVIDUAL EXPRESSION OF MY LIFE and also in looking at the object, have had the individual pleasure of realising that my personality was objective, visible to the senses and thus a power raised beyond all doubt."*

physician-turned-philosopher Raymond Tallis in his remarkable recent book on *The Hand*[12] where, like Derrida, he also uses this idea as the basis for a theory of the emergence of human consciousness. The book elaborates on Friedrich Engels' famous statement that

*"[t]he hand is not only the organ of labour, it is also the product of labour."*[13]

Tallis suggests that out of the "objectifying" of human action in the repeated patterns of technical processes and the material forms of tools and artefacts, emerges a growing awareness of the hand itself as a kind of proto-technical object.

While this might also, perhaps, explain the special prominence given to the image of the hand in many examples of Palaeolithic cave painting, the major implication of this is that the ability to see one's actions "sedimented" in the solid residues of technical practices might even have been the stimulus to sensing the subject-object status of one's own body: that is what Merleau-Ponty has described as our curiously ambiguous status as integrated "body-subjects". Therefore, moving beyond the idea of the technological prosthetic, which was introduced at the beginning of this discussion, rather than thinking of technology merely as an extension of the body, it may even be true to say that thinking of ourselves as having a body, and having a choice as to what to do with it, might actually be a consequence of our prehistorical development of technology.

And perhaps this is what Karl Marx also had in mind when he described the satisfaction of the manual worker in the contemplation of the completed task:

*"Supposing we had produced in a human manner; each of us would in his production have doubly affirmed himself and his fellow men. I would have objectified in my production my individuality and its peculiarity and thus both in my activity enjoyed an INDIVIDUAL EXPRESSION OF MY LIFE and also in looking at the object have had the individual pleasure of realising that my personality was objective, visible to the senses and thus a power raised beyond all doubt."*[14]

Marx's statement also highlights two complementary forms of creative experience, which seem to result from the process of making he is describing: firstly the experience of the maker in taking up and transforming a raw material into an object of use; and secondly the experience of the user in taking up an object consciously shaped for human interaction. So, if it is true that we produce ourselves as subjects in the creative action of producing objects, I would argue that we also continually reproduce ourselves as creative subjects in the act of taking up and using objects. The symmetry between the process of constructing and both inhabiting and interpreting architecture,[15] which this suggests, is what I want to try and illustrate in the final section of the chapter.

**Construction and Occupation**

On one level, most of what has been said above is simply a reminder that technology is, after all, "much more important than we think". But, in an attempt to go beyond this, I would also like to suggest that the same kind of body-brain feedback loop still contributes to our understanding of architecture in an everyday sense. For example, I would argue that we "read" an environment on two distinctly different levels, both in terms of how it might have been made and also how it might be used. This experience relies partly on a building's capacity to record and reveal both the processes of construction and inhabitation: including the tectonic articulation of the materials and the processes of its making, and also the accumulating traces of occupation left by the users' patterns and habits of use. In other words, to paraphrase the words of the philosopher Paul Ricoeur, we might say that the hermeneutics of architecture involves a double process of interpretation: both of the "space behind the work" (understanding the intentions of its author/designer) and, more importantly, the space "in front of the work", understanding the experience the building makes possible for its future "readers"/occupiers.[16]

One recent example of what I would like to call "symmetry" between construction and occupation comes from the New Art Gallery in Walsall, UK, designed by Caruso St John Architects and opened in the year 2000. On the upper landings of the main stair lobbies a number of footprints are clearly visible, fragments of builders' boot-marks cast permanently into the concrete floor.

While the power-float machines used to finish the floors would normally be expected to smooth these over, in this case the architects have perhaps even encouraged the builders not to be too careful about "covering their tracks". As these traces of the construction process appear alongside the more transient footprints left by the building's users, they invite us, in a modest way, to link how the building was made and how it might be occupied. If there is a cognitive function behind this connection between the tectonics of construction and the "tectonics of use", then perhaps it is also best explained by the words of Paul Ricoeur:

*"... it must be said that we understand ourselves only by the long detour of the signs of humanity deposited in cultural works."*[17]

**Jonathan Hale**
is associate professor and reader in architectural theory at the University of Nottingham, where he serves as head of the Architectural Humanities Research Group. He is the author of a series of books including: *Merleau-Ponty for Architects* (in press), and co-editor of *Rethinking Technology: A Reader in Architectural Theory* (2007), and he holds a PhD from the University of Pennsylvania School of Design.

Caruso St John Architects, New Art Gallery Walsall (2000), Footprints cast into the polished concrete floor of the stair landing. © J.Hale, 2006.

Caruso St John, New Art Gallery Walsall (2000), Concrete floors and stairs. © J.Hale, 2006.

Caruso St John, New Art Gallery Walsall (2000), Concrete wall and timber paneling. © J.Hale, 2006.

1 __ Tim Ingold, "Tool-Use, Sociality and Intelligence," *Tools, in Language and Cognition in Human Evolution*, ed. Kathleen Gibson and Tim Ingold (Cambridge: Cambridge University Press, 1993), p. 433.

2 __ Maurice Merleau-Ponty, *Phenomenology of Perception* (Abingdon: Routledge, 2012 [1945]), p. 153.

3 __ Brian Massumi, "Stelarc: The Evolutionary Alchemy of Reason," in *The Virtual Dimension: Architecture, Representation and Crash Culture*, ed. John Beckmann (New York: Princeton Architectural Press, 1998), p. 336.

4 __ Andy Clark and David Chalmers, "The Extended Mind," *Analysis* 58 (1998), pp. 7-19.

5 __ Bernard Stiegler, *Technics and Time, 1: The Fault of Epimetheus* (Stanford, California: Stanford University Press, 1998), p. 152.

6 __ Steven Mithen, *The Prehistory of Mind: A Search for the Origins of Art, Religion and Science* (London: Thames and Hudson, 1996), p. 7.

7 __ André Leroi-Gourhan, *Gesture and Speech* (Cambridge, MA: MIT Press, 1993 [1964]), p. 82.

8 __ Michael C. Corballis, *From Hand to Mouth: The Origins of Language* (Princeton, New Jersey: Princeton University Press, 2002).

9 __ Ingold, "Tool-Use, Sociality and Intelligence," in Kathleen Gibson and Tim Ingold (eds.), *Tools, Language and Cognition in Human Evolution* (Cambridge: Cambridge University Press, 1993), pp. 429-45.

10 __ Jacques Derrida, *Of Grammatology*, trans. Gayatri Chakravorty Spivak (Baltimore, MD: Johns Hopkins University Press, 1976), p. 84.

11 __ Merleau-Ponty, *Phenomenology of Perception*, pp. 130-148; Marcel Mauss, "Techniques of the Body," trans. Ben Brewster, in *Techniques, Technology and Civilisation*, ed. Nathan Schlanger (New York: Berghahn Books/Durkheim Press, 2006[1935]), pp. 77-95.

12 __ Raymond Tallis, *The Hand: A Philosophical Enquiry into Human Being* (Edinburgh: Edinburgh University Press, 2003)

13 __ Friedrich Engels, *Dialectics of Nature*, trans. Clemens Dutt (London: Lawrence and Wishart, (1940 [1882]), p. 281.

14 __ David McLellan, *The Thought of Karl Marx* (London: Papermac, 1995), p. 23.

15 __ According to Marco Frascari, *Monsters of Architecture: Anthropomorphism in Architectural Theory* (New York: Rowman and Littlefield, 1991), p. 170.

16 __ Paul Ricoeur, *Hermeneutics and the Human Sciences: Essays on Language, Action, and Interpretation*, trans. John B. Thompson (Cambridge: Cambridge University Press, 1981), p. 141.

17 __ Ricoeur, *Hermeneutics and the Human Sciences*, p. 143.

# Acknowledgements

A book of this nature builds upon a collective interest and effort. Due to its attention to tectonics, sustainability and building culture, it naturally must draw upon an accumulation of widespread knowledge coming from particular research environments and well-established reflective practices. During our research on the topics and concepts behind *Towards an Ecology of Tectonics* we have been fortunate to include insights from a number of people and furthermore been able to engage some of them as co-authors of this book.

First and foremost we would like to thank our co-authors – a rare group of highly respected practicing architects and academics whom we consider to be some of the finest advocates as well as critical thinkers dealing with the complex topic of the book. Experts on different topics across practice and theory, they were speakers at two international symposiums held in the spring and fall of 2012. In random order we would like to acknowledge: Børre Skodvin, partner of Jensen & Skodvin Architects in Oslo; Johan Celsing, founder of Celsing Architects in Stockholm; Peter Thorsen, partner and CEO of Lundgaard & Tranberg Architects in Copenhagen; and Bijoy Jain, founder of Studio Mumbai in Mumbai; David Leatherbarrow, professor of architecture, University of Pennsylvania School of Design; Fredrik Nilsson, professor and head of the department of architecture, Chalmers University of Technology; and Jonathan Hale, associate professor in architectural theory and deputy head of the Architecture + Urbanism Research Division, University of Nottingham.

The various contributions to this book originate from a number of activities that have been generated by the collaborative scientific project "Towards a Tectonic Sustainable Building Culture", in the period from 2010–2014. The project has included the three primary research institutions in the architectural field in Denmark: The Royal Danish Academy of Fine Arts, Schools of Architecture, Design and Conservation – School of Architecture (KADK), Aarhus School of Architecture (AAA) and Danish Building Research Institute, Aalborg University (SBI/AAU). Individuals and small groups in these independent academic environments have conducted research on the complex topics of this book for a number of years. However, it was not until we defined a collaborative research project and were awarded a sizeable research grant from The Danish Council for Independent Research | Humanities in 2010 that we were able to join forces and to study the research field from a purely architectural perspective – by use of a holistic or rather a "polyhistorical" approach, which we believe is fundamental and is needed more than ever in contemporary architectural research.

Further support was also provided by a generous grant donated by the VELUX Visiting Professor Programme in the period from 2011–2012, which made it possible for us to invite architect Bijoy Jain, founder of Studio Mumbai, to take part in the project as a sort of external reviewer. Bijoy Jain is a profound thinker and has formed a unique architectural practice based on the process of learning through making, putting into practice architectural and material studies through large-scale mock-ups, traditional skills, local building techniques, and the ingenuity that arises from limited resources. Ongoing exchanges of ideas and deep discussions with Bijoy Jain about the field of tectonics have been of great value for the project as a whole.

We'd like to thank the following persons, who participated as opponents in the second symposium of 2012: Natalie Mossin, president of the Danish Architectural Association; Lars Juel Thiis, partner of CUBO Architects, Aarhus; and Søren Nielsen, partner of Tegnestuen Vandkunsten, Copenhagen, for their fruitful comments and true engagement in this ongoing discussion. Also, we want to acknowledge our colleagues from a number of Scandinavian universities and architectural practices with whom we have exchanged ideas over the years, an exchange that has led to the idea of forming the Scandinavian Tectonics Network. Among these are: Marie Frier Hvejsel, assistant professor, Aalborg University; Tim Anstey, associate professor and director of research, KTH, Stockholm; Marja Lundgren, PhD student and partner of White Architects, Stockholm; Finn Hakonsen, associate professor, NTNU, Trondheim; Anne Sigrid Nordby, environmental coordinator, Asplan Viak A/S; Michael U. Hensel, professor of architecture and director of RCAT – Research Center for Architecture and Tectonics, AHO, Oslo; and Marius Nygaard, professor at the Institute for Architecture, AHO, Oslo. Also, Børre Skodvin and Fredrik Nilsson, mentioned previously, are considered part of this network.

Colleagues and students from our various research environments and teaching programs have throughout the years generously shared their ideas and taken part in central discussions that have been of great inspiration and guidance.

For supporting us with critical reviews on our first and second (successful) research grant application we'd like to thank the following: Kasper Sanchéz Vibæk, former associate professor at CINARK – Centre for Industrialized Architecture, KADK; and Jacob Kristoffer Hansen, head of the research division, KADK. We'd like to thank additional colleagues at the Institute of Architectural Technology, KADK – Nini Leimand, associate professor; Mette Jerl Jensen, research assistant; Anne-Mette Manelius, senior consultant, Construction & Management, Danish Technology Institute; and Johannes Rauff Greisen, consultant, Concrete Division, Danish Technology Institute – for insisting on challenging the traditional understanding of tectonics in their PhD dissertations prior or parallel to this present research project.

Also, we are very grateful to our dedicated project secretary, M.Arch. Tenna Beck, who has supported us and coordinated our work and doings during the years of the project, and Senior Clerk Birthe Færch, who has meticulously taken care of the project accounts.

Finally, we want to thank the Royal Danish Academy of Fine Arts Schools of Architecture, Design & Conservation – School of Architecture for hosting the project and supporting the publication of this book.

# Endnotes and Credits

**Introduction**
**An Ecology of Tectonics**
by Anne Beim p. 20-23

1 __ Paraphrase that sums up the hypothesis of the collaborative Danish research project "Towards a Tectonic Sustainable Building Practice", 2010-2014. The Royal Danish Academy of Fine Arts School of Architecture, Aarhus School of Architecture, and The Danish Building Research Institute (SBI)/ AAU.

2 __ Predictions and analyses offered by the International Energy Agency show that international political ambitions to decrease the global emissions of GHG might have failed; therefore, new political agreements and strategies must be settled. *World Energy Outlook Special Report 2013: Redrawing the Energy Climate Map, Edition 2013, 132 pages*. www.worldenergyoutlook.org/ energyclimatemap (01062014).

3 __ In the late '60s Elias Cornell, the renowned Swedish professor of architectural history and theory at KTH (The Royal Technical University of Sweden), wrote one of the first comprehensive books on the "history of building technology", in which he describes its evolutionary trajectory as part of human civilisation based on a holistic approach. Cornell, Elias, *Byggnadstekniken; Metoder och idéer genom tiderna* (Stockholm: Byggförlaget, 1970). James Strike, a British architectural historian, also follows a similar line of thought in his book *Construction into Design: The Influence of New Methods of Construction on Architectural Design 1690-1990* (Oxford: Butterworth Architecture, 1991). Finally, the scholars William W. Braham and Jonathan A. Hale have collected a number of central texts that include historical, philosophical and sociological subjects dealing with technology and practical and ethical dimensions of change, development and evolution in architecture. William W. Braham and Jonathan A. Hale (eds.), *Rethinking Technology: A Reader in Architectural Theory* (London and New York: Routledge, Taylor & Francis Group, 2007).

4 __ A critical analysis of the contemporary construction industry and implications for architectural design and the profession is offered: Stephen Kieran and James Timberlake, *Refabricating Architecture: How Manufacturing Methodologies Are Poised to Transform Building Construction* (New York: McGraw-Hill, 2004)

5 __ E.g., certifications such as LEED, BREEAM and DGNB have now spread worldwide, influencing local building legislation and policies.

6 __ William W. Braham and Jonathan A. Hale (eds.), *Rethinking Technology: A Reader in Architectural Theory* (London & New York: Routledge, Taylor & Francis Group, 2007), xiii.

7 __ Ursula M. Franklin, *The Real World of Technology* (Toronto: House of Anansi Press, 1999), 95.

8 __ Jan S. Kauschen, findings of the PhD Project: *Sustainable Integrated Product Deliveries in New Building and Renovations Projects*, The Royal Danish Academy of Fine Arts School of Architecture.

9 __ *The Energy-efficient Buildings PPP: research for low energy consumption buildings in the EU*, European Commission, 2013, http://ec.europa.eu/ research/press/2013/pdf/ppp/ eeb_factsheet.pdf, (01062014).

10 __ To be certified as a Passive House, a maximum of 15 kWh/m$^2$ must be used for heating the building space. In addition, the total primary energy consumption (domestic hot water, space heating, ventilation, pumps, household facilities, lights, etc.) may not exceed 120 kWh/m$^2$. Finally, the airtightness of the building may not exceed 0.6 m$^3$/h$^3$/m$^3$, in a blower door test. http://www.passiv.de/ en/02_informations/02_passive-house-requirements/02_passive-house-requirements.htm

11 __ Marco Frascari, *Monsters of Architecture: Anthropomorphism in Architectural Theory* (Lanham, MD: Rowman & Littlefield Publishers Inc., 1996), 117.

12 __ Opening statement from the International Symposium, *Tectonic Thinking and Practice in Architecture*, 2012 held at The Royal Danish Academy of Fine Arts - School of Architecture in Copenhagen.

13 __ Elias Cornell, *Byggnadstekniken; Metoder och idéer genom tiderna* (Stockholm: Byggförlaget, 1970), 9-30.

14 __ Marcus Vitruvius Pollio (first century BCE), Leon Battista Alberti (1404-1472), and Andrea Palladio (1508-1580); Rowland, I.D. & Howe, T.N. (eds.), *Vitruvius. Ten Books on Architecture* (Cambridge: Cambridge University Press, 1999); Leon Battista Alberti, *On the Art of Building in Ten Books*, transl. by Joseph Rykwert, Neil Leach and Robert Tavernor (Cambridge, Massachusetts: MIT Press, 1988/92); Palladio, Andrea, *The Four Books on Architecture*, transl. by Robert Tavernor and Richard Schofield (Cambridge, Massachusetts: MIT Press, 1997).

15 __ Karl Bötticher, *Die Tektonik der Hellenen*, vols. 1-2, (Potsdam: Riegel, 1844-52).

16 __ Gottfried Semper, *Der Stil in den technischen und tektonischen Künsten; oder, Praktische Aesthetik: Ein Handbuch für Techniker, Künstler und Kunstfreunde*, vol. 2 (Frankfurt am Main: Verlag für Kunst & Wissenschaft, 1860).

17 __ Gottfried Semper, *Die vier Elemente der Baukunst: Ein Beitrag zur vergleichenden Baukunde*, (Braunschweig: Friedrich Vieweg und Sohn, 1885).

18 __ Kenneth Frampton, *Studies in Tectonic Culture: The Poetics of Construction in the Nineteenth and Twentieth Century Architecture* (Cambridge, Massachusetts: MIT Press, 1995), 4.

19 __ Detlef Mertins, "Walter Benjamin and the Tectonic Unconscious: Using Architecture as an Optical Instrument", in: *Modernity Unbound. Architecture Words 7*, (London: Architectural Association, 2011), 119.

20 __ Detlef Mertins, *Modernity Unbound. Architecture Words 7*, 119.

21 __ Gottfried Semper, *Der Stil in den technischen und tektonischen Künsten;*

*oder, Praktische Aesthetik: Ein Handbuch für Techniker, Künstler und Kunstfreunde*, vol. 2 (Frankfurt am Main: Verlag für Kunst & Wissenschaft, 1860), 107.

22 __ Gevork Hartoonian, *Ontology of Construction: On Nihilism of Technology in Theories of Modern Architecture* (New York: Cambridge University Press, 1994), 29-30.

23 __ Marco Frascari, "Tell-the-tale Detail", in: *VIA*, No. 7 (Philadelphia: University of Pennsylvania, 1984).

24 __ Hartoonian, *Ontology of Construction: On Nihilism of Technology in Theories of Modern Architecture*, 11.

25 __ Hartoonian, *Ontology of Construction: On Nihilism of Technology in Theories of Modern Architecture*, 20-21.

26 __ Mari Hvattum, *Gottfried Semper and the Problem of Historicism* (Cambridge: Cambridge University Press, 2004), 79-80.

27 __ Hvattum, *Gottfried Semper and the Problem of Historicism*, 82-83.

28 __ Neil Leach, David Turnbull & Chris Williams, (2004), *Digital Tectonics* (Chichester: John Wiley & Sons Ltd, 2004); Jesse Reiser + Nanako Umemoto, *Atlas of Novel Tectonics* (New York: Princeton Architectural Press, 2006); Stephen Kieran & James Timberlake, *Refabricating Architecture: How Manufacturing Methodologies Are Poised to Transform Building Construction* (New York: McGraw-Hill Companies, 2004).

29 __ Neil Leach, David Turnbull & Chris Williams, (2004), *Digital Tectonics*, (Chichester: John Wiley & Sons Ltd, 2004), 5-6.

30 __ Jesse Reiser + Nanako Umemoto, *Atlas of Novel Tectonics*, Plate 66.

31 __ Stephen Kieran & James Timberlake, *Refabricating Architecture*, xii.

**An Etymology of Tectonics**
by Karl Christensen p. 26-27


__ *Dansk-nygræsk ordbog*, Rolf Hesse, Munksgaard, 1995.
__ ΣΥΓΧΡΟΝΟ Ελληνοδανικό Δεξικό, Rolf Hesse, ΕΚΔΟΣΕΙΣ ΠΑΤΑΚΗ, nygræsk-dansk ordbog, 1997.
__ *Oldgræsk Dansk*, Carl Berg, Gyldendal, 2003.
__ *Classical Greek Dictionary*, Oxford, 2002.
__ *Græsk-Dansk ordbog – til brug for den studerende ungdom*, Paul Arnesen, Gyldendalske Boghandlings Forlag, 1830.
__ *Spørgsmålet om teknikken – og andre skrifter*, Martin Heidegger, Gyldendal, 1957/1999.
__ *Kunstværkets oprindelse*, Martin Heidegger, Gyldendal, 1950/1994.
__ *Politikens filosofileksikon*, Politikens Forlag, 1983.
__ *Dansk Fremmedordbog*, Munksgaards Ordbøger, 1997.


**Photosuite opening the book**

__ p. 4
House Stoneman – Exterior Killeenaran, County Galway, Ireland. Architect: Richard Murphy Architects. Completed: 1997. © CINARK

__ p. 5
House Stoneman – Living room Killeenaran, County Galway, Ireland. Architect: Richard Murphy Architects. Completed: 1997. © CINARK

__ p. 6+7
Social Housing Mulhouse, Mulhouse, France. Architect: Lacaton & Vassal Architects. Completed: 2005. © CINARK

__ p. 8
Bruno Mathssons Summer House, Frösakull, Halland County, Sweden. Architect: Bruno Mathsson. Completed: 1960. © CINARK

__ p. 9
Vesthimmerlands Museum, Aars, Himmerland County, Denmark. Architect: Per Kirkeby & Jens Bertelsen. Completed: 1999. © Ole Meyer

__ p. 10
Letterfrack Furniture College – Wood workshop, Letterfrack, County Galway, Ireland. Architect: O'Donnell + Tuomey Architects. Completed: 2001. © CINARK

__ p. 11
Transformation of Housing Block Paris, Tour Bois le Prêtre, Paris, France. Architect: Lacaton & Vassal Architects. Completed: 2011. © CINARK

__ p. 12
Ricola Europe, Mulhouse, Brunstatt, France. Architect: Herzog & de Meuron. Completed: 1993. © CINARK

__ p. 13
Brookfield Youth and Community Centre, West Tallaght, Dublin, Ireland. Architect: Hassett-Ducatez Architects. Completed: 2008. © CINARK

__ p. 14+15
Centro de Estudios Hidrográficos, Madrid, Spain. Architect: Michel Fisac. Completed: 1963. © CINARK

__ p. 16
McCormick Tribune Campus Center, Chicago, Illinois, USA. Architect: OMA. Completed: 2003. © CINARK

__ p. 17
Fireplace for Children, Skjermveien Kindergarten, Trondheim, Norway. Architect: Haugen / Zohar Arkitekter. Completed: 2009. © CINARK

__ p. 24+25
Trim Castle, Trim, Co. Meath, Ireland. Architect: Unknown. Completed in the 12th century. © Thomas Bo Jensen

**Towards an Ecology of Tectonics**
*The Need for Rethinking Construction in Architecture*


ISBN 978-936681-86-4



Editors:
Anne Beim
Ulrik Stylsvig Madsen

Editorial Board:
Charlotte Bundgaard
Karl Christiansen
Thomas Bo Jensen

Proofreading:
Ilze Mueller

Graphic design:
Søren Damstedt, Trefold

Printing and binding: Graspo CZ, a.s., Zlín, Czech Republic